D1064764

Functional Behavior of Orthopedic Biomaterials

Volume I: Fundamentals

Editors

Paul Ducheyne, Ph.D.
Associate Professor of
Biomedical Engineering
University of Pennsylvania
Philadelphia, Pennsylvania

Garth W. Hastings, D.Sc., Ph.D., C.Chem., F.R.S.C.
Head of Biomedical Engineering Unit
North Staffordshire Polytechnic, and
Honorary Scientific Officer
North Staffordshire Area Health District
Medical Institute
Hartshill, Stoke-on-Trent
England

CRC Series in Structure-Property Relationships of Biomaterials

Series Editors-in-Chief

Garth W. Hastings, D.Sc., Ph.D., C.Chem., F.R.S.C.

Paul Ducheyne, Ph.D.

CRC Press, Inc.
Boca Raton, Florida

Library of Congress Cataloging in Publication Data
Main entry under title:

Functional behavior of orthopedic biomaterials.

(CRC series in structure-property relationships
of biomaterials)
Includes bibliographies and index.
Contents: v. 1. Fundamentals — v. 2. Applications.
1. Orthopedic implants — Materials. 2. Orthopedic
implants — Materials — Corrosion. I. Ducheyne, Paul.
II. Hastings, Garth W. III. Series. [DNLM: Bio-
compatible materials — Therapeutic use. 2. Bone and
bones — Surgery. 3. Implants, Artificial. WE 190 F979]
RD755.5.F86 1984   617'.307   83-3737
ISBN 0-8493-6265-2 (v. 1)
ISBN 0-8493-6266-0 (v. 2)

Direct all inquiries to CRC Press, Inc., 2000 Corporate Blvd., N.W., Boca Raton, Florida, 33431.

© 1984 by CRC Press, Inc.

International Standard Book Number 0-8493-6265-2 (Volume I)
International Standard Book Number 0-8493-6266-0 (Volume II)

Library of Congress Card Number 83-3737
Printed in the United States

SERIES PREFACE

Biomaterials science is concerned with surgical implants and medical devices and their interaction with the tissues they contact. Their study, therefore, includes not only the properties of the materials from which they are made, but also those of the tissues which will accept them. Metals, ceramics, and macromolecules are the artifacts. Bone tendons, skin, nerves, and muscles are among the tissues studied. Prosthetic materials, implants, dental materials, dressings, extra corporeal devices, encapsulants, and orthoses are included among the applications.

It is not only the materials *per se* which interest the biomaterials scientist, but also the interactions in vivo, because it is at the interface between implant and tissues that the success of a procedure will be decided. This approach has led to the concept of a more aggressive role for biomaterials in the actual treatment of disease. Macromolecular drug delivery systems are receiving considerable attention, especially those with the capacity for targeting specific sites in the body. Sensing and control of body processes is a logical extension of this. There is much to be done before these newer developments become established.

The science of biomaterials has grown and developed over the last few years to become an accepted discipline of study. It is opportune, therefore, to systematize the study of biomaterials in order to improve their application in medical science, since that is the end point of all studies. That is the aim of this series of books on *Structure-Property Relationships in Biomaterials*. Knowledge of structure and the influence on properties is fundamental to any materials science study; it is a more complex problem to obtain the knowledge from tissue materials, as the living organism has a great capacity for change and adaptation in response to a stimulus. The stimulus may be chemical, electrical, or mechanical. The biomaterials scientist endeavors to identify and to use these stimuli and responses to improve the in vivo acceptability of the materials.

Many institutions and agencies have promoted the science of biomaterials. Societies now exist for this purpose. The Biological Engineering Society (U.K.) founded in 1960 formed a Biomaterials Group in 1974. In the same year the Society for Biomaterials was founded in the U.S. The European Society for Biomaterials (1976) was followed by Canadian and Japanese Societies (1979). All societies play a major role in disseminating knowledge through conferences and publications.

This series is complementary to these society activities. It is hoped that it will not only provide a basis of knowledge, but also its own stimulus for further progress. The series is inevitably selective. In part this is due to the editors' choice, in part to the availability of authors. The editors wish to thank those who fulfilled their agreements. Without them this series would not have been possible.

G. W. Hastings
Series-Editor-in-Chief

PREFACE

Once the properties of metals, ceramics, and polymers are known and the biological tissues have been characterized, the next step in the study of biomaterials is the interaction between them. Those interactions can influence the integrity of the implant material as well as cause changes in surrounding tissues or distant organs. The ultimate answer to the efficacy of any given implant material can therefore only be obtained when that material has a functional size and shape and is clinically used. Although the importance of the clinical trial step is stressed here, the previous stages of biomaterials evaluation such as the in vitro experiments and the animal testing must not be overlooked (this is, for instance, discussed in the volumes, *Metal and Ceramic Biomaterials*, of this same series).

The present volume is the first of two volumes that describes some of the interactions between orthopedic implant materials and the human environment. The first chapter of this volume describes in detail what we understand by "functional behavior". In addition, it outlines important areas of development and problems that would profit from a greater research effort.

The other chapters of this volume provide data on the properties of the osseous tissues. In these tissues many orthopedic implants are hosted. Carter and Wright (Chapters 2 and 3) describe the ultimate strength of cortical bone and the macroscopic directionality of properties. Those data are essential in the studies dealing with the response of bone tissues and bone structures on external loading stimuli.

Huiskes (Chapter 4) discusses the various methods that are available to analyze the stresses and strains in bones, joints, and implant assemblies. Thus, this chapter deals with the static response to loading stimuli. In contrast, Van der Perre (Chapter 5) discusses the dynamic response; the major parts of that chapter are, first, the fundamental concepts of dynamic response and, second, the dynamic behavior of bone tissues, bone structures, and bone-implant assemblies.

Included in *Functional Behavior of Orthopedic Biomaterials, Volume II: Applications*, are Part I, which assesses the fixation and status of total joint replacements, and Part II, which presents an evaluation of wear behavior, acrylic bone cement, and fixation materials under functional conditions. Included is an analysis of design, geometry, and stresses in artificial joints.

It is obvious that in books of this nature it is impossible to obtain a comprehensive coverage of all subjects that are relevant to its title. Rather a selection of important topics can be proposed. Furthermore, however meaningful the title and its topics could be, the eventual value of the book is to a large extent determined by the value of the separate contributions. Therefore the editors wish to thank all the authors for their hard work, without which these books would never have been published. We hope that their efforts will be rewarded by illuminating the way of all those in the field or those wishing to join it.

P. Ducheyne and G. W. Hastings
January 1983

THE EDITORS

Garth W. Hastings, D.Sc., Ph.D., C.Chem., F.R.S.C., is a graduate of the University of Birmingham, England with a B.Sc. in Chemistry (1953) and a Ph.D. (1956) for a thesis on ultrasonic degradation of polymers. After working for the Ministry of Aviation he became Senior Lecturer in Polymer Science at the University of New South Wales, Sydney, Australia (1961 to 1972). During this time he was Visiting Professor at Twente Technological University, Enschede, The Netherlands (1968-69), advising on their program in biomedical engineering. While in Australia, he became associated with Bernard Bloch, F.R.C.S., Orthopedic Surgeon, Sydney Hospital, and began a fruitful collaboration in the uses of plastics materials in surgery.

In 1972 he returned to England as Principal Lecturer in the Biomedical Engineering Unit of the North Staffordshire Polytechnic and the (now) North Staffordshire Health District with responsibility for research. With a particular interest in biomaterials research his own work has encompassed carbon fiber composites for surgical implants, adhesives, bioceramics, prosthesis performance in vivo, and electrical phenomena in bone. He is a member of British and International Standards Committees dealing with surgical implants and of other professional and scientific bodies, including Companion Fellow of the British Orthopaedic Association and Editor of the international Journal *Biomaterials*. He was elected President of the Biological Engineering Society in the U.K. (B.E.S.) in October, 1982. He was awarded a D.Sc. from the University of Birmingham in 1980 for a thesis in the field of biomedical applications of polymers. He has recently been appointed Acting Head of the department.

Paul Ducheyne, Ph.D. obtained the degree of metallurgical engineering from the Katholieke Universiteit Leuven, Belgium, in 1972. Subsequently he worked at the same university towards a Ph.D. on the thesis "Metallic Orthopaedic Implants with a Porous Coating" (1976). He stayed one year at the University of Florida as an International Postdoctoral N.I.H. Fellow and a CRB Honorary Fellow of the Belgian-American Educational Foundation. Thereafter he returned to the Katholieke Universiteit Leuven. There he was a lecturer and a research associate, affiliated with the National Foundation for Scientific Research of Belgium (NFWO). He recently joined the University of Pennsylvania, Philadelphia, as an Associate Professor of Biomedical Engineering and Orthopedic Surgery Research.

Dr. Ducheyne has published in major international journals on mechanical properties and design of prostheses, porous materials, bioglass, hydroxyapatite, and microstructural methods of analysis of biomedical materials. He is member of the editorial board of *Biomaterials, Journal of the Engineering Alumni of the University of Leuven, Journal Biomedical Materials Research,* and *Journal Biomechanics and Comtex System for Biomechanics and Bioengineering.*

He became active in various societies and institutions and has held or is holding the positions of Chairman-Founder of the "Biomedical Engineering and Health Care Group" of the Belgian Engineering Society, Secretary of the European Society for Biomaterials and member of the Board of Directors of Meditek (Belgian Institution to promote biomedical industrial activity).

CONTRIBUTORS

Dennis R. Carter, Ph.D.
Associate Professor
Department of Mechanical Engineering
Stanford University
Stanford, California

Ian C. Clarke, Ph.D.
Associate Professor of Orthopedics and
  Mechanical Engineering
University of Southern California
Los Angeles, California

Peter Griss, M.D.
Professor Doctor Medicine
Orthopedic Clinic
Lindenhof
Manheim, West Germany

Rik Huiskes, Ph.D.
Associate Professor
Laboratory of Experimental Orthopedics
University of Nijmegen
 and Consultant
Department of Mechanical Engineering
Eindhoven University of Technology
Eindhoven, The Netherlands

John C. Keller, Ph.D.
Department of Biophysical Dentistry
Medical University of South Carolina
Charleston, South Carolina

Eugene P. Lautenschlager, Ph.D.
Professor
Department of Biological Materials
Northwestern University
Chicago, Illinois

Harry A. McKellop, M.Sc.
Instructor in Orthopedic Research
University of Southern California Medi-
  cal School
Los Angeles, California

Samuel I. Stupp, Ph.D.
Assistant Professor of Bioengineering
  and Ceramic Engineering
Department of Ceramic Engineering
Polymer Group
Bioengineering Program
University of Illinois at Urbana-
  Champaign
Urbana, Illinois

Georges Van der Perre, Ph.D.
Professor
Analytical Mechanics and Biomechanics
Katholieke Universiteit
Leuven, Belgium

Timothy M. Wright, Ph.D.
Associate Professor
Cornell University Medical College
Ithaca, New York and
Associate Scientist
The Hospital for Special Surgery
Department of Biomechanics
New York, New York

FUNCTIONAL BEHAVIOR OF ORTHOPEDIC BIOMATERIALS

Volume I: Fundamentals

Functional Behavior of Orthopedic Biomaterials
Yield Characteristics of Cortical Bone
Macroscopic Directionality in Bone
Principles and Methods of Solid Biomechanics
Dynamic Analysis of Human Bones

Volume II: Applications

Part I: Assessment of Total Joint Replacement

The Fixation of Permanent Implants: A Functional Assessment
Assessment of the Clinical Status of Total Joint Replacement

Part II: Evaluation of Materials and Devices

Evolution and Evaluation of Materials-Screening Machines and
Joint Simulators in Predicting In Vivo Wear Phenomena
Structure and Properties of Acrylic Bone Cement
Design, Fixation, and Stress Analysis of Permanent Orthopedic
Implants: The Hip Joint
Biological Fixation of Implants

TABLE OF CONTENTS

Volume I

Chapter 1
Functional Behavior of Orthopedic Biomaterials 1
P. Ducheyne and G. W. Hastings

Chapter 2
Yield Characteristics of Cortical Bone.. 9
D. R. Carter and T. M. Wright

Chapter 3
Macroscopic Directionality in Bone 37
T. M. Wright and D. R. Carter

Chapter 4
Principles and Methods of Solid Biomechanics... 51
R. Huiskes

Chapter 5
Dynamic Analysis of Human Bones.. 99
G. Van der Perre

Index 161

Chapter 1

FUNCTIONAL BEHAVIOR OF ORTHOPEDIC BIOMATERIALS

P. Ducheyne and G. W. Hastings

TABLE OF CONTENTS

I.   Functional Behavior: Definition 2

II.   General Requirements of Orthopedic Biomaterials 2
    A.   Biocompatibility 3
    B.   Sufficient Mechanical Properties 3
    C.   Low Friction and Wear 3
    D.   Dimensions Appropriate to Its Location 3
    E.   Long-Term Functionality 3
    F.   Possibility to be Sterilized 3

III.   Testing Protocol for Implants and Implant Materials 3

IV.   The Dichotomy Between Experimental and Clinical Assessment 4
    A.   Materials Development 4
    B.   Fixation Materials 5
    C.   In Vitro Test Methods 5
    D.   Biomechanical Techniques 5
    E.   Animal Experimentation 6
    F.   Clinical Evaluation 6

V.   Limitations in Orthopedic Biomaterials Development 6

VI.   Conclusion 7

Reference 7

I. FUNCTIONAL BEHAVIOR: DEFINITION

This volume of the series in *Structure-Property Relationships of Biomaterials* deals with orthopedic biomaterials. The scope of this volume is to review the information available to evaluate properly the use of materials for orthopedic surgical use. In previous volumes, a comprehensive treatment was given of the existing biomaterials. Metals, ceramics, glasses, carbons, polymers, and composites were described. Their bulk and surface properties which are of particular relevance to use as an implant or as part of an implanted device were discussed. In addition the tissues of the musculoskeletal system were analyzed. The data presented in those volumes now form the basis for assessing these materials in actual implant form. Not only is it important to study on their own the properties of biomaterials as materials; it is of equal value to consider these materials in size and shape appropriate for implantation and to study the various interactions which can occur among the different materials and the physiological environment. The implant material is part of a whole structure, and its properties can only be fully assessed if it is known how the structure as a whole functions. This basic and fundamental concept is what we call ''functional behavior''.

The example of total hip replacement arthroplasty (THR) may be of help in more clearly delineating the meaning of this concept. THR is performed in an attempt to alleviate pain, reconstruct the joint, and restore normal motion at the hip joint. There are a number of procedures possible, but we can basically look at THR as an operation where, as it stands today in the majority of cases, four types of material interact with one another: an ultrahigh molecular weight polyethylene (UHMWPE) acetabular cup, a metal femoral component, an orthopedic polymethylmethacrylate bone cement, and surrounding tissues. The eventual clinical success of the THR procedure is dependent upon the fixation of the components, the absence of infection which can be foreign material mediated, the stability of the calcar which can be influenced by cytotoxic reactions on wear and abrasion particles or by strain shielding by the stiff metal stem, the mechanical performance of the stem, the stem-cement junction, and the stability of the osseous tissue building. This list of examples is clearly not exhaustive. It is merely intended to show the extent to which the function of the THR is dependent on the different materials and their interactions. Summarizing these input data for the evaluation of artificial hip joint functionality, there is

1.   The synthetic material with its own mechanical, physical, and chemical surface and bulk properties
2.   The geometry of these materials
3.   The effect of a particular type of loading (This loading can vary from patient to patient)
4.   The effect of various conditions of body fluids on the chemical and mechanical physical properties of the implant materials
5.   The properties of the tissues considering that these properties may be directional and that the tissues contain fluids, organic material, and mineral components

These data are used to describe the biomechanical, biochemical, and possibly bioelectrically triggered processes basically underlying the stability of the artificial hip joint and the overall state of health of the patient.

II. GENERAL REQUIREMENTS OF ORTHOPEDIC BIOMATERIALS

In terms of the implant material, there are generally some six requirements considered for successful biomaterial usage. It depends upon the particular type of application which requirement stands out as the more important one. The requirements of some implants and implant materials will be described in detail in the several chapters of this book. A brief account of the general requirements follows here.

A. Biocompatibility

The term biocompatibility covers a two-way phenomenon associated only with the material itself: implant materials may not be adversely affected by the physiological environment, and the local or remote tissues and organs may not be harmed by the presence of the material. This definition does not include mechanical or geometrical effects. This requirement is the most fundamental since any biomaterial comes into interaction with the physiological environment. Adverse effects on the implant material are, for instance, corrosion of metal and ceramic implants or degradation of polymers by the saline solution of the body. Adverse local effects are necrosis or resorption of tissue, unfavorable cellular reactions, and synergistic action with bacteria to cause infection. Possible systemic reactions are hypersensitivity, toxicity, and carcinogenicity.

B. Sufficient Mechanical Properties

In many uses of biomaterials, especially those in orthopedics, the implants are structural parts subjected to loads which can be high, of cyclic nature, and at different strain rates. Mechanical properties, such as yield and ultimate strength, ductility, modulus of elasticity, endurance limit, and viscoelastic behavior, are therefore important characteristics.

C. Low Friction and Wear

As a result of increasing usage of artificial joints, considerable work has been performed to develop the best combination of materials for the gliding surfaces: a low-friction coefficient and a high-wear resistance are paramount. The minute particles released from the joint surfaces can elicit tissue reactions interfering with the overall joint performance. An artificial joint made by metal against itself is not suited for internal use as a surgical implant.

D. Dimensions Appropriate to its Location

Prostheses must be of such form that they can be implanted without causing undue damage. As an example, a knee prosthesis is mostly made of such a form that only a minimal amount of tissues must be excised.

E. Long-Term Functionality

In certain applications, a controlled surface or bulk reaction of the biomaterial is a beneficial effect one wishes to obtain. Functionality here is defined as fulfilling its intended function over the whole lifetime of the patient for a permanent implant or until its purpose is achieved. As an example, under appropriate conditions bioglasses form a bond to bone as a result of a sequence of reactions at the surface. Hydroxylapatite and tricalciumphosphates can be bioresorbable; this means that they are digested or transported through surrounding cells. The rate of reaction of these bioactive materials should not impair the function to which the implants are intended, thus the rate of reactivity has to be closely controlled.

F. Possibility to be Sterilized

Implants can be sterilized by several methods, including autoclaving (steam at 120 to 140°C), gamma radiation, or sterilization by ethylene oxide. Depending upon the material, one of these methods is used which does not adversely affect the material properties.

III. TESTING PROTOCOL FOR IMPLANTS AND IMPLANT MATERIALS

There are in general five consecutive stages of evaluation which are successfully executed, before any material or design is widely used for clinical purposes:

1.   The evaluation by simple in vitro tests

2.   The evaluation by complex in vitro tests
3.   The experimentation in animals: biocompatibility tests
4.   The evaluation in animals as a functional implant
5.   The clinical trial step

When considering a new material as a gliding material in a joint replacement, the protocol described above would first typically be the use of simple wear test methods to compare the behavior of the new material with existing materials; if successful this material may then be fabricated in the form of an implant and tested in a joint simulator; at the same time, one may start implantation of this material in unloaded conditions in bulk and in powder form. Again if these steps are satisfactory, one may proceed to the next step which is animal experimentation with the newly developed material in prosthesis form; this is to test its behavior as a gliding material in vivo. If sufficient information is gathered showing that the new device or material has substantial potential for successful, hazard-free clinical performance, a clinical trial step is initiated.

IV. THE DICHOTOMY BETWEEN EXPERIMENTAL AND CLINICAL ASSESSMENT

Depending upon the type of implant material or device, some of the five steps may be redundant. However, it is obvious that successfully carrying through experimental programs to implement new materials and devices requires substantial expenditure. It may therefore be tempting to reduce costs as much as possible. This can be achieved by cutting expenses in each step of a proper protocol and by moving quickly and eagerly to human experimentation. In addition to cost saving, a rationale for doing so is provided by the general and specific lack of comparability between the in vitro and animal experiments on one side and the clinical situation on the other side. For instance, it is well known that some animal models may invariably yield positive results, while human implantation under similar circumstances is bound to failure. There is thus a dichotomy between experimental laboratory programs and human implementation. This dichotomy represents a major challenge to the advancement of understanding the behavior of implants and implant materials. As an example, to understand the fate of metal ion release from metal implants, careful in vitro and animal corrosion studies checked for their relevance to the clinical reality, are needed.

Fast and uncontrolled progression to human use may be queried not just on the basis of ethical concepts, but also by scientific criteria. A closed loop research outline with built-in feedback is indicated to improve progressively the properties and availability of materials and devices. The necessity of excellent research programs, predictive of the fate of materials and devices in clinical reality, becomes even more obligatory considering the ever-increasing time of implantation for which present implants are designed. Fifteen years of life in service may presently be expected. At present we may observe the following areas where careful experimental protocols predictive for long-term clinical situations are very valuable and/or urgently needed.

A. Materials Development

1.   Metals with high fatigue limit are being developed for use as the femoral component of stemmed-type total hip prostheses.
2.   Ceramics are tested by fracture mechanics methods in order to study the possible degradation by the physiological fluids as a function of time.
3.   Composite materials are becoming available for a number of applications. An advantage which is hoped will be used to its full benefit is the reduced modulus of elasticity. Attention is being paid to the effect of the directionality of the properties and to the

interfaces between constituents. These interfaces may well constitute the weak links of the system.

4. Surface-coated materials have received little attention up to now. This is deplorable, since by careful selection of materials for bulk and surface of an implant, a near optimal combination of properties may be obtained. The technology for applying coatings is far more advanced in other engineering fields than in biomedical engineering.

B. Fixation Materials

1. Porous coatings can be made from ceramic, polymer, composite, or metal base materials; these porous structures allow bony ingrowth and thus stable fixation of permanent implants provided there is initial stability and sufficient pore size. Biocompatibility, stress pattern in and around the implant, enhancement of bone ingrowth, and manufacturing methods to obtain sufficient substrate strength and design aspects are being considered for optimal implant functioning.
2. The mechanical, elastic, histological, immunological, and biochemical aspects of bioglasses, hydroxyapatite, and tricalciumphosphates are being scrutinized in order to evaluate predictively the bonding with tissues and the fate of the dissolving ions.

C. In Vitro Test Methods

1. The metals which are used are in the human body in their passive state. Even in this condition, metal ions are released into the physiological environment. However, little is known about the mechanisms of this release, the form under which the metallic elements are subsequently found in the tissues, the transport mechanism to distant organs, and the effect on systemic functions. The onset of a better understanding of these fundamental questions can now be perceived. In vitro corrosion tests, in the presence of suitable concentrations of enzymes, proteins, and lipoproteins, which induce similar effects to those observed in vivo, are urgently needed.
2. New materials are developed for artificial joints, but a reproducible, standardized wear testing procedure which can predict the wear for a service life up to 10 to 20 years is, however, not available.
3. Manufacturers have spent great time and effort to produce metals with high fatigue strength. However, a simple reproducible fatigue testing method to test the fatigue properties of the actual implant shape has not yet been produced, although several are being discussed in national and international standards bodies.

D. Biomechanical Techniques

1. There has been much debate on the value of finite element models for stress analysis of biological structures with implants. It appears essential that some allowance is made for the discontinuous displacement actually occurring at the interfaces in implant systems. In addition if a two-dimensional method is used, the properties of the elements must be carefully selected in order that the model yields realistic results.
2. In vitro tests on the kinematics and stresses of implants are easier to reproduce when instead of cadaveric bones, synthetic material is used to replace the bones in the testing. Some materials, such as glass fiber epoxies and phenolic resin fiber composites, have been proposed.
3. The mechanical properties of tissues surrounding the implants have mostly been measured on fresh ex vivo specimens. The process of preparing these specimens may, however, induce changes in the properties. A full understanding of these changes has not yet been reached.

4.   The loading pattern varies significantly from patient to patient. Engineering analyses to provide statistical data on load patterns at the major joints under pathological or replacement arthroplasty conditions would be useful.

E. Animal Experimentation

1.   Animal models to evaluate long-term functionality of implants and implant materials are very costly. In addition, problems in interpretation arise because of some dissimilarities: there may be differences in the physiology of the test animal and man; it is also difficult to reproduce in the animals the pathological conditions for which implants are developed; in addition, animal management is largely different from patient treatments. In the field of orthopedic biomaterials, there are few generally accepted protocols for animal experimentations, unlike other fields.
2.   The techniques for assessing the interface between implants and tissues are still being improved. This holds both for the mechanical and for the histological evaluation.

F. Clinical Evaluation

1.   Eventually new materials are clinically tried. Scientific assessment of the results is greatly facilitated when the clinical trial is confined to ''a one-parameter study''. Anticipated influential parameters must be kept constant except for the property of the material under study.
2.   Clear and uniformly accepted definitions of clinical success are needed. A decisive conclusion on the value of radiolucency at the implant-bone interface with regard to prospective evaluation of total joint replacement arthroplasty would be valuable.
3.   Implant retrieval and analysis helps to establish correlations between laboratory results and clinical reality.

V. LIMITATIONS IN ORTHOPEDIC BIOMATERIALS DEVELOPMENT

Few materials can presently be used for temporary and permanent implants, especially when the implant is subjected to high stresses. This lack of a suitable choice of materials frequently hampers the design of new devices. Efforts are made to develop materials for specific medical uses, but experimental and clinical limitations slow down the pace at which new materials become available. Hench[1] discerned six experimental limitations:

1.   The required research effort is one at the forefront of both biological and materials sciences.
2.   The instrumental tools for the required analyses have only become available within the last few years.
3.   The application of these instruments necessitates considerable modifications of the instrumental operating methods.
4.   The maintenance of the interface in preparation for analysis is a major variable.
5.   The selection of appropriate animal model implant configuration and evaluative test protocols is very complex.
6.   Production of a sufficient number of materials in any shape and size necessary for full evaluation is very difficult.

Once human evaluation has started, there are also clinical limitations:

1.   The major one is related to the ill-defined quantitative measures for success. Clinical

observations are difficult to assess quantitatively. The main symptom for total joint replacement surgery being pain, how can pain relief after surgery be evaluated quantitatively? How can the degree of symptom recurrence be measured? It is possible that functionality can be assessed more quantitatively by measuring angles of rotation between different segments of the skeleton, but it still cannot be compared to thorough engineering methods which positively discriminate one condition from another. In addition there is the interference of the patient's psychology, e.g., by expressing his opinion on pain level and thus personal degree of satisfaction.

2.   Since current tendencies in orthopedic implant applications are towards longer implantation periods, increasingly longer clinical observation periods become necessary. However, this is very difficult since there have continuously been changes and improvements in device design, operative technique, and postoperative treatment. Furthermore, the largest number of implants has been implanted during approximately the last 5 years.

3.   New developments are mostly evaluated at centers of excellence. Subsequent general usage means putting the materials and devices into the hands of surgeons who are less experienced with the technique and may reach a lower level of achievement.

4.   There may be conflicting requirements between the intended goal of the implant and the operative procedure. Total joint replacements are intended to restore functionality at the joint, but if biological attachment is used, some immobilization is required.

VI. CONCLUSION

The aspects involved in biomedical materials research are so different that projects are conducted by multidisciplinary teams, including material scientists, mechanical engineers, histopathologists, and surgeons. Transfer of knowledge among team members of different disciplines is then paramount. Were past failures of implants related to communication gaps existing among the different professional areas of this field? Implants failed due to wrong design, inappropriate materials choice, deficient production, faulty operative procedure, or incorrect patient management. The knowledge transfer gap probably does not exist anymore in well-established groups. However, just as important, there might still be considerable delay in the transfer of new understanding in the medical disciplines or the engineering sciences to the other profession. A total of 400 types of different knee prostheses currently exist, indicating a high degree of unsatisfactory performance of knee implants as yet. However, the centers where the present clinical experience, the mechanical analyses of the knee joint, and the availability of improved materials result in a fool-proof design are very few.

It is therefore the aim of this volume to present the necessary details for generally defining and assessing any orthopedic material problem. The different chapters are chosen in such a way that they may provide a basic understanding of the various aspects of orthopedic biomaterials. At the same time, they give a full account of the state-of-the-art for the more important areas of development. The electrical effects in osseous tissue associated with some orthopedic procedures are not discussed, since it was felt that this is too far beyond the scope of materials' analysis.

REFERENCE

1. **Hench, L. L. and Clark, A. E.,** Adhesion to bone, in *Biocompatibility of Orthopedic Implants*, Vol. 2, Williams, D. F., Ed., CRC Press, Boca Raton, Fla., 1982, 129.

Chapter 2

YIELD CHARACTERISTICS OF CORTICAL BONE

Dennis R. Carter and Timothy M. Wright

TABLE OF CONTENTS

I.   Introduction 10

II.   Stress-Strain Relationships in Monotonic Loading to Fracture 10

III.   The Influence of Bone Composition and Aging 12

IV.   Microyield Phenomena 15

V.   Fatigue Damage Accumulation and Failure 17

VI.   Acoustic Emission Studies 20

VII.   Microscopic Examination of Yield and Fatigue Damage 21

References 35

I. INTRODUCTION

Many of the initial mechanical testing studies of bone were directed at determining the ultimate strength of whole bones and the material strength of bone tissue under different loading conditions. In recent years, there has been an increased interest in the mechanical and micromechanical events in bone tissue which occur during yield or fatigue loading, prior to catastrophic failure. The micromechanical events associated with yield and fatigue are of fundamental mechanical interest in understanding the material behavior of bone tissue. In addition, however, these events may influence the normal biological remodeling processes which determine bone deposition and resorption and ultimately the mechanical integrity of the whole bone. In this chapter, the basic stress-strain behavior of bone tissue as determined in standard materials testing procedures will be reviewed. Other investigations which have attempted to reveal the microyield and fatigue behavior of bone tissue will then be introduced. Finally, studies illustrating the microscopic nature of bone tissue yielding will be presented.

II. STRESS-STRAIN RELATIONSHIPS IN MONOTONIC LOADING TO FRACTURE

The studies by Reilly et al.[1] and Reilly and Burstein[2] on the tensile and compressive behavior of cortical bone at different orientations are the most complete examinations of the elastic and ultimate properties of human cortical bone in the literature. Reilly et al.[1] conducted tensile and compressive mechanical tests of bone specimens oriented with the long bone axis. A rapid strain rate of approximately 0.02 to 0.05/sec was used for all testing. The resulting stress-strain curves showed that bone behaves in a manner similar to other engineering materials. Stress-strain curves in tension and compression consist of an initial elastic region which is nearly linear. This region is followed by yielding. Considerable nonelastic ("plastic") deformation occurs prior to ultimate failure (Figure 1). The nonelastic region of the curve reflects diffuse, irreversible microdamage introduced throughout the bone structure. Although this region is often referred to as the plastic portion of the stress-strain curve, the micromechanical events responsible for this nonelastic behavior in bone are distinctly different from those observed during plastic deformation of metals. The conventions used for establishing yield stress, ultimate stress, yield strain, ultimate strain, elastic modulus, and "plastic modulus" are shown in Figure 1.

Reilly and Burstein[2] presented the first systematic investigation of the anisotropic properties of cortical bone using macroscopic strain measurements. Testing was conducted at strain rates between 0.02 and 0.05/sec. Specimens were extracted from a population ranging in age from 19 to 80 years. Testing was done under fully wet conditions, and extensometers attached directly to the bone surface were used to measure bone strain during testing. All specimens tested had a secondary Haversian microstructure. The mean value of the longitudinal modulus was found to be approximately 50% greater than that of the transverse modulus. The ultimate strength was greater in compression than in tension for specimens oriented in both the longitudinal and transverse directions. Specimens loaded in the transverse direction were significantly weaker in both tension and compression than the longitudinally oriented specimens. In addition, the transverse specimens tended to fail in a more brittle manner, with little nonelastic deformation subsequent to yielding.

Since bone is a viscoelastic material, the stress-strain curves depicted in Figure 1 are influenced by the applied strain rate during mechanical testing. The ultimate strength and the elastic modulus of specimens tested in a direction along the whole bone axis approximately double as the strain rate is increased from 0.001 to 10.0/sec. In addition, the yield strain and ultimate strain tend to decrease with increasing strain rate.[3-5]

Evans and Lebow[6] and others[7-9] have pointed out that the material characteristics of bone

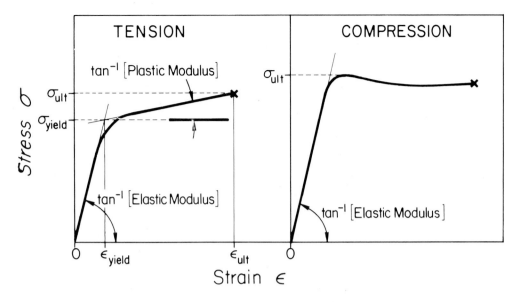

FIGURE 1. Stylized stress-strain curves of human cortical bone specimens tested in tension and compression. (From Carter, D. R. and Spengler, D. M., *Clin. Orthop.*, 135, 192, 1978. With permission.)

tissue measured in the laboratory are sensitive to the procedures followed in the preparation and testing of the specimens. These considerations are particularly important in assessing the yield characteristics of bone. Drying of bone prior to or during testing causes the tissue to behave in a more brittle manner. Due to the many differences in specimen preparation procedures, it is therefore not surprising to find some disagreement in the literature concerning the yield and nonelastic portions of the curves shown in Figure 1. Earlier studies wherein care was not exercised in maintaining bone wetness report stress-strain curves with little additional deformation after yield.

Burstein et al.[7] pointed out that in the literature there are consistent differences between the ultimate strength properties measured using axial loading and those measured under bending loads. To reconcile these apparent discrepancies, an elastic-perfectly plastic mathematical model for the bending behavior of bone specimens was presented. The model was used to analyze specimens with both circular and rectangular cross sections exposed to three point bending mechanical tests. The underlying concept of their approach was that the reported bending strength of bone tissue (modulus of rupture) is calculated assuming a linear distribution of stress and strain through the specimen at failure. If significant yielding occurs prior to failure, however, the assumption of a linear stress distribution is no longer valid (Figure 2). The bending strength calculations would then seriously overestimate the stresses present in the extreme fibers of the bending specimens at failure.

The results of their analyses indicate that the differences in bone axial and bending strength reported in the literature can be quantitatively explained if one accounts for the plastic behavior of bone tissue. The elastic-perfectly plastic model provides a reasonable first-order correction to the bending strengths calculated using the linear elastic beam theory formula. For example, using simple beam theory, the modulus of rupture of a circular cross section specimen was determined as 379 ± 29 MPa at a strain rate of about 0.1/sec. The elastic-perfectly plastic model would predict extreme fiber stresses of 195 ± 15 MPa for such specimens. These calculated extreme fiber stresses correlate reasonably well with the uniaxial tensile strength of 172 ± 21 MPa obtained experimentally.

Burstein and co-workers saw two significant implications from their observations of yielding in bone tissue. Firstly, they asserted that there must certainly be incidents of in

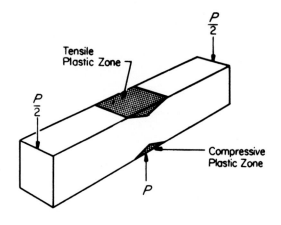

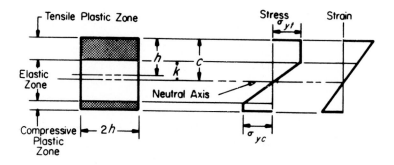

FIGURE 2.   Stress and strain profiles for a bending load on a specimen with a square cross-section, assuming unequal tensile and compressive yielding. (From Burstein, A. H., Currey, J. D., Frankel, V. H., and Reilly, D. T., *J. Biomech.*, 5, 35, 1972. With permission.)

vivo loading wherein bone tissue yields, yet gross fracture does not occur. It then becomes important to understand the influence of such yield phenomena on the subsequent biological behavior of the tissue. Secondly, they pointed out the importance of yielding in determining the bending strength characteristics of whole bones. Their calculations show that due to yielding, the bending strength of long bones is approximately twice as great as that which could be expected if bone behaved as a brittle material.

III. THE INFLUENCE OF BONE COMPOSITION AND AGING

Burstein et al.[10] examined the in vitro contribution of collagen and mineral to the elastic-plastic properties of bovine Haversian cortical bone. Progressive surface decalcification of bone specimens with dilute hydrochloric acid produced progressive decreases in the tensile yield stress and ultimate stress, but caused no change in the yield strain or ultimate strain unless decalcification was completed (Figure 3). The slope of the nonelastic region of the stress-strain curve remained constant throughout the decalcification process. The authors stated that these findings are consistent with an elastic-perfectly plastic model for the mineral phase of bone tissue in the presence of the organic matrix. They further suggested that the mineral contributes the major portion of the tensile yield strength, while the magnitude of the plastic modulus is a function only of the properties of collagen, which itself plays a minor role in the tensile yield strength of bone.

Burstein and associates found that the degree of mineralization has a rather strong influence

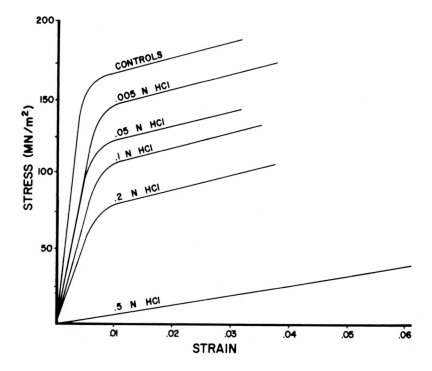

FIGURE 3. Average stress-strain curves for specimens of bovine bone treated to create increasing degrees of decalcification. (From Burstein, A. H., Zika, J. M., Heiple, K. G., and Klein, L., *J. Bone Jt. Surg.*, 57a, 956, 1975. With permission.)

on modulus and yield stress, but has little influence on yield strain. This finding suggests that bone tissue yielding might be better described by strain criteria than by stress criteria. The importance of tissue strain as a yield criterion was advanced by Currey.[11] This view is also supported by work on pathologic bone conducted by other researchers. Bell and associates[12] investigated the influence of a diet supplement consisting of a lathrogen, β-aminoproprio-nitrile (BAPN), on the bending properties of dry rat femora. BAPN is now known to inhibit the activity of the enzyme lysyl oxidase, which is essential for the formation of aldehyde side chains needed for collagen cross-links.[13,14] The results of their study showed that BAPN produced a significant decrease in the stiffness and strength of the femora of young, growing rats. The loss of bone properties with BAPN treatment was evident even after the data were adjusted to compensate for differences in whole long bone geometry. The estimated yield strain was not affected by BAPN treatment, however, the ultimate strain at failure was significantly increased. The addition of thyroid hormones to the rat diet inhibited the loss of material properties caused by BAPN treatment, but produced no significant changes in the material properties of the femora of normal rats.

Spengler and co-workers[15] investigated the influence of the in vivo administration of BAPN on the torsional properties of whole rat femora. In addition to impaired bone collagen cross-linking, a 5% decrease in bone ash was observed.[16] The bones were maintained in a fully wet condition prior to and during mechanical testing. Torsional stiffness and strength of the experimental femora were less than those of the control femora, however, the total deformation to failure was 47% greater. Yield deformations were not reported. No change in bone geometry was produced by BAPN treatment at the doses used.

Weir et al.[17] examined the influences of the rachitogenic diet on the bending properties of whole rat femora. To facilitate estimations of the bone material properties, beam theory was used to compensate for the influence of differences in geometry between control and experimental bones. The bones from rats on the rachitogenic diet were significantly weaker and had a lower modulus of elasticity than bones from normal rats. In addition, the control bones showed significantly less ultimate deformation to failure. Deformation at yield was not reported. The average ash content was 59% for the normal rats and 37% for the experimental rats. The loss of ash content and mechanical properties caused by the rachitogenic diet was significantly less when the diet was supplemented with vitamin D.

Aging causes significant changes in collagen cross-linking and mineralization in cortical bone. Rasmussen et al.[18] and Bailey[19] demonstrated that with aging the collagen fibers in connective tissue gradually increased in stability. These age-dependent changes in collagen have generally been attributed to an increase in the intermolecular covalent cross-links between molecules.[20] Fujii et al.[21] reported that the amount of soluble collagen in human bone decreases from about 1.9 to approximately 0.2% between 3 to 18 years of age. The amount of soluble collagen continues to decrease very slowly to 0.1% at 89 years. These data reflect a progressive increase in collagen cross-linking and are coincidental with changes in bone mineralization reported by other investigators. Currey and Butler[22] showed that the percentage of bone ash in human cortical bone increases dramatically between 2 to 17 years of age, but changes very little between 26 to 48 years of age. Black et al.[23] reported a small increase in bone ash between 20 to 80 years of age.

Aging is also associated with significant changes in the morphology and composition of cortical bone. The morphologic changes are caused primarily by internal bone remodeling throughout life. Kerley[24] examined the microscopic, age-related changes that occur in human cortical bone by counting the number of osteons, osteon fragments, and non-Haversian canals and estimating the percentage of circumferential lamellar bone in the outer third of the cortex in ground sections from the midshafts of 126 human femurs, tibias, and fibulas. Age correlations with these microstructural parameters were made throughout an age range from birth to 95 years. Increasing age was associated with a loss in the percentage of area occupied by the circumferential lamellar bone. In the cross sections examined, the number of osteons and the number of osteon fragments were also found to correlate positively with age. These findings can be directly related to the internal remodeling of bone which occurs progressively throughout life.

Mechanical properties of bone tissue in children and adults were investigated by Vinz[25,26] and Currey and Butler.[22] Vinz studied the tensile properties of bone specimens extracted from humans 0 weeks to 85 years of age. A progressive increase in mineralization was demonstrated for specimens throughout this age range. The tensile strength and modulus of elasticity were found to increase up to the age of 40 years. The ultimate strain decreased with age. The study of Currey and Butler was based on femoral cortical bone specimens from 18 subjects between 2 to 48 years of age. Specimens were loaded in three-point bending to failure. Results were consistent with those of Vinz. The bone from children had a lower modulus of elasticity, bending strength, and ash content than the adult bone specimens. However, this bone demonstrated an increased amount of total deformation to failure and also absorbed more energy before fracture (Figure 4). The authors emphasized that a large region of nonelastic deformation is observed in children's bone. The large postyield region in this bone is responsible for the large amount of energy absorbed prior to fracture.

Burstein et al.[27] examined mechanical properties of cortical bone specimens extracted from adult human femurs and tibias. Mechanical testing was done in tension, torsion, and compression for specimens from a population ranging in age from 21 to 86 years. No significant differences were found between the mechanical properties of male and female bone specimens. The tibial specimens had greater ultimate strength, stiffness, and ultimate strain than the femoral specimens. Consistent decreases with age for ultimate strength, ultimate strain, yield strength, and elastic modulus were found in the femoral, but not in the tibial specimens. No influence of age on yield strain was reported. No statistically

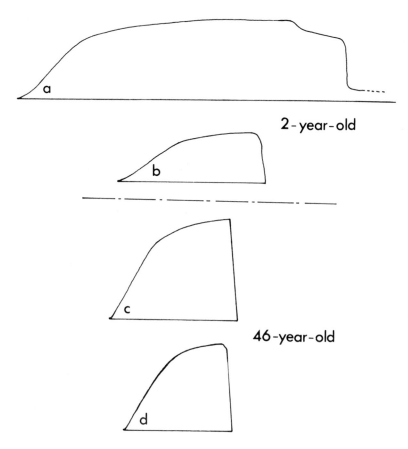

FIGURE 4.   Load-displacement curves of bone specimens loaded in bending. Curves a and b show the greatest and least deformation of specimens from 2-year-old bone. Curves c and d are comparable curves from 46-year-old bone. (From Currey, J. D. and Butler, G., *J. Bone Jt. Surg.*, 57a, 810, 1975. With permission.)

significant differences were observed in the tensile properties of specimens from normal, osteoporotic, and corticosteroid-treated individuals. The lack of statistical significance in these tests, however, may be attributed to the small number of specimens tested. Femoral bone specimens were shown to have an increased plastic modulus with age. Burstein and co-workers suggested that the increase in slope of the stress-strain curve in the postyield region may be caused by structural changes in bone collagen which are age-dependent. Burstein and associates suggested that from a structural point of view the most important age change that occurs in bone tissue is the decrease in ultimate strain at failure. The decrease in ultimate strain in the tensile tests was approximately 5% per age decade in femoral bone and 7% per decade in tibial bone. This decrease in ultimate strain was responsible for the 32 and 42% decreases in tensile energy absorption in the femoral and tibial tissue respectively between the 3rd and 9th decades of life.

IV. MICROYIELD PHENOMENA

Monotonic stress-strain characteristics of cortical bone can conveniently be discussed in terms of the elastic, yield, and nonelastic (or plastic) regions. Careful examination of the elastic portion of the curves reveal that this region is, in fact, slightly nonlinear and that a small degree of strain softening occurs with increasing load. To investigate this phenomena,

Bonfield and Li[28] conducted mechanical tests to determine the microscopic yield stress of bovine femoral and tibial bone specimens. Microscopic yield stress is defined as that stress which produces an offset strain of 2×10^{-6} mm/mm from linearity. The strain offsets from linearity were referred to as plastic strain by Bonfield and Li. Strain was measured using a Tuckerman optical strain gage with mechanical contact on the specimen. All tensile specimens were oriented in the longitudinal whole bone direction. A load-unload technique was used in which a series of increasing tensile stresses were applied at a very low strain rate of 3.3×10^{-4}/sec. The total strain (elastic and plastic) and the residual plastic strain after unloading were measured as a function of time.

Bonfield and Li determined that the average microscopic yield stress for 10 specimens were approximately 3 MPa. This is an extremely low stress level which would be exceeded by bone tissue during everyday in vivo activities.[29,30] As the stresses were increased beyond the microscopic yield stress, the plastic strains progressively increased.

The plastic strain produced in these tests was largely anelastic strain which was recovered after unloading within a period of about 10 min. However, some nonrecoverable, permanent strain was measured in the specimens. The percentage of recoverable plastic strain was higher at higher loading levels. For example, specimens loaded and unloaded at stress levels of 31, 62, and 93 MPa had recoverable anelastic strains of 44, 63, and 67%, respectively, of the initially recorded plastic strain.

Bonfield and Li also found that, on repeating a test to the same stress level, the percentage of recoverable strain was considerably larger in cycles subsequent to the first loading cycle. This uniqueness of the first loading cycle in bone has been noted by others and was extensively discussed by Black.[31]

A more recent study by Bonfield and O'Connor[32] further investigated the anelastic deformation properties of bone. Femoral and tibial bone specimens were machined from bovine and rabbit long bones. Tensile loading-unloading tests were conducted in a manner similar to that used by Bonfield and Li.[28] However, to measure strain, foil strain gages were bonded directly to the bone specimen. Their tests revealed a microscopic yield stress of 12 ± 8 MPa. The departure from linear elastic behavior is depicted in Figure 5. At low stress levels (Curve A), the specimens behaved as a linearly elastic material. When the elastic limit was first exceeded, the loading and unloading curves coincided only at the maximum and zero stress levels (Curve B). At higher stress levels, the loading and unloading curves did not coincide at zero stress on unloading, giving initial residual, nonelastic strain (i.e., an open hysteresis loop), which recovered with time to form a closed hysteresis loop (Curve C). For loading and unloading curves up to a strain of 0.003 mm/mm, they found that all of the nonelastic strain was recovered after unloading if one waited a long enough time (maximum of 40 min). From these results, they concluded that the nonelastic strain is entirely anelastic. They also introduced the concept of a friction stress which defines the onset of anelastic deformation.

The results of Bonfield and O'Connor[32] appear to contradict the earlier results of Bonfield and Li[28,33] which revealed nonelastic strains which were nonrecoverable after the microscopic yield stress was exceeded. Bonfield[62] has suggested that this discrepancy may be related to the relative "wetness" of the specimens tested. In the later studies, in which all strains were recoverable, greater care was taken to prevent specimen drying.

Ascenzi and Bonucci[34-36] have conducted extensive micromechanical studies on the behavior of single osteons extracted from cortical bone tissue. The samples were tested in a microtesting machine to failure in tension and compression. Load and strain were monitored continuously. Osteon strain was measured with a microwave micrometer during testing. Their results indicate that the stress-strain relationships in osteons are strongly influenced by the orientation of collagen fiber bundles. The yield and ultimate strains of osteons tested in both tension and compression decreased with increasing calcification and the yield and

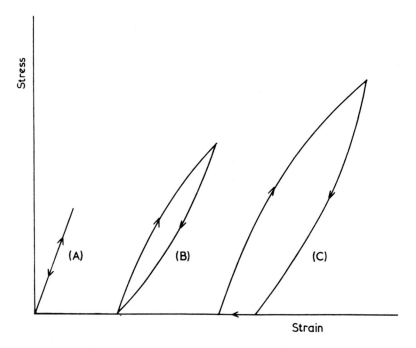

FIGURE 5.   Schematic illustration of the three types of load-unload cycles observed by Bonfield and O'Connor. (A) A straight line (linear elastic); (B) a closed hysteresis loop (anelastic); (C) an initially open hysteresis loop which closes with time at zero stress (also anelastic). (From Bonfield, W. and O'Connor, P., *J. Material Sci.*, 13, 202, 1978. With permission.)

ultimate stresses increased. For the compression tests, failure was initiated by the appearance of microscopic fissures on planes of high shear stress.

Ascenzi and Bonucci[36] conducted tensile tests of fully calcified osteon samples with alternating lamellae which were extracted from human femoral shafts. Careful examination of the stress-strain curve from each osteon showed one or more distinct bends in the curve at relatively loss stresses (Figure 6). To interpret this nonlinear behavior, they compared the osteons to fiber-reinforced plastic materials called cross-ply laminates which behave in a similar manner. They also examined the damage induced in the osteon microstructure using electron microscopy. Based on their findings, they suggested that the change in slope at low stresses is due to a failure of the interfibrillar cementing substance in those lamellae having fiber bundles oriented perpendicularly to the loading direction, and to yielding at the canaliculi previously filled with osteocyte processes. Their results tend to support the hypothesis that nonrecoverable, permanent strains are introduced to bone tissue at relatively low stress levels. They pointed out that their findings contradict the generally held opinion that under normal in vivo activity, the proportional limit is never reached.

V. FATIGUE DAMAGE ACCUMULATION AND FAILURE

Before attempting to characterize the nature of fatigue damage in bone, it is reasonable to examine the fatigue behavior of other materials. Although metals and composite materials fail in fatigue, the fatigue process in metals is quite different from that of composites. In metals, repeated loading causes the accumulation of plastic strains and associated plastic slip lines which lead to fatigue crack initiation.[37] Fatigue cracks propagate under repeated loading until final catastrophic failure occurs. Metals subjected to repeated loading do not exhibit a significant loss of stiffness or ultimate strength until after the crack is initiated.[38,39]

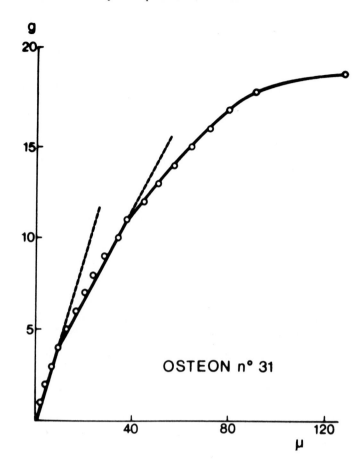

FIGURE 6.   Tensile load-displacement curve recorded for a fully calcified
osteon with alternating fibril orientation in adjacent lamellae. Two changes in
slope are shown at low load levels. (From Ascenzi, A. and Bonucci, E., *J.
Biomech.*, 9, 65, 1976. With permission.)

Composite materials fail in fatigue by widely distributed damage processes which may include
delamination, matrix crazing, fiber failure, void growth, and matrix cracking.[38] As a result,
a composite material often exhibits a more gradual and progressive loss of stiffness and
strength throughout its fatigue life.[38-41]

The fatigue behavior of a material is related to its behavior under a single applied loading
to failure.[38] Compact bone specimens loaded to failure in tension were shown by Burstein
and co-workers to exhibit a stress-strain curve consisting of an elastic region followed by
yield, a flatter postyield regime, and failure.[6,7,42,43] Metals and some composite materials
exhibit similar loading curves. However, the micromechanical events at yield and postyield
for metals are very different from those of composites. Yielding in metals is caused by
plastic flow and is accompanied by the formation of plastic slip lines. A metal specimen
loaded into its postyield regime, unloaded and subsequently reloaded, exhibits little or no
loss of stiffness and no loss of ultimate strength. In composite materials, the postyield regime
is caused by multiple damage modes, such as microcracking, debonding, void growth, and
fiber breakages. A composite material loaded into the postyield regime, unloaded, and
subsequently reloaded typically exhibits a loss of stiffness and ultimate strength. Insights
into the mechanical nature of yielding in a material may therefore be gained by studying its
behavior under repetitive loading.

Carter and Hayes[44] investigated the residual strength and stiffness of bone specimens that

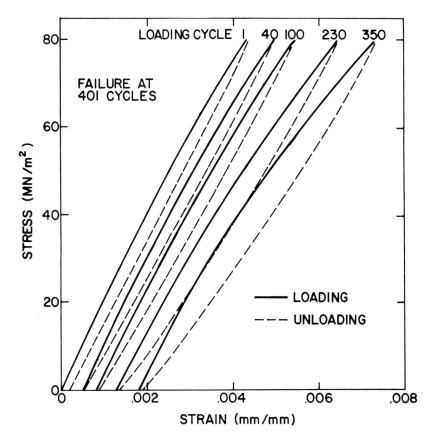

FIGURE 7.   Cyclic stress-strain curves recorded during tensile fatigue of cortical bone specimen. (From Carter, D. R. and Hayes, W. C., *J. Biomech.*, 10, 325, 1977. With permission.)

were subjected to repeated cyclic loading. Residual strength tests consisted of rotating bending fatigue loading followed by tensile tests to failure. Specimens loaded for approximately half the number of cycles required for fatigue fracture exhibited a significant 13% loss of tensile strength compared to those specimens which were not fatigued. Specimens tested under tensile fatigue loading exhibited a gradual and progressive loss of stiffness which in some specimens exceeded 20% when failure occurred. The loss of bone stiffness was accompanied by increasingly nonlinear loading behavior and an increase in hysteresis (Figure 7). Similar fatigue behavior observed in concrete and composite materials is attributed to cumulative microcracking, debonding, void growth, and fiber breakage.[38,45] These findings strongly suggest that the postyield regime of compact bone is the result of diffuse structural damage.

Compact bone in compression is stronger and exhibits a smaller postyield regime than in tension.[42,46] This behavior in monotonic loading is similar to that of glass fiber reinforced cement (GRC), a composite of two brittle materials.[40] Fatigue tests of GRC in tension cause a reduction in modulus which is strikingly similar to that seen with the bone specimens. Allen[40] noted that the GRC modulus reduction in fatigue is caused by cement cracking. Although there can be no direct physical analogy drawn between bone and GRC, the comparison of mechanical properties illustrates that one need not rely only on the concepts of plastic flow or crazing to explain the nonelastic behavior of bone.

Carter et al.[47,48] conducted uniaxial fatigue tests of cortical bone specimens machined from adult human femora. The specimens were tested under strain control with total strain

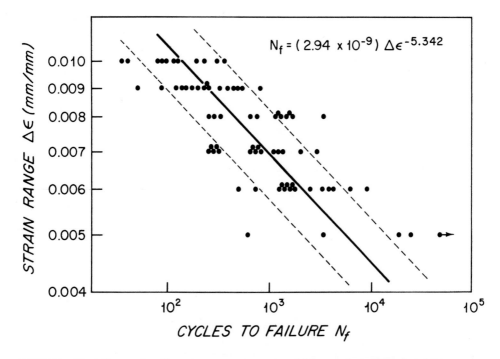

$$N_f = (2.94 \times 10^{-9}) \Delta\epsilon^{-5.342}$$

FIGURE 8.   The influence of cyclic strain range on the number of fatigue cycles to failure for adult human bone specimens. From Carter, D. R., Caler, W. E., Spengler, D. M., and Frankel, V. H., *J. Biomech.*, 14, 461, 1981. With permission.)

ranges from 0.005 to 0.010 mm/mm. The influence of mean strains of 0.002 (tension), 0.0, and −0.002 (compression) was studied. A cyclic strain rate of 0.01/sec was used, resulting in loading frequencies between 0.5 and 1.0 Hz. The results were consistent with previous findings that bone fatigue is a gradual damage process accompanied by a progressive increase in hysteresis and a loss of bone stiffness. The total number of cycles to fatigue failure was influenced only by the total strain range and was not affected by mean strain. Bone was shown to have extremely poor fatigue resistance. Fully reversed cyclic loading to one half of the yield strain (± 0.0034 mm/mm) caused fatigue fracture in approximately 1000 cycles (Figure 8).

The monotonic (single-loading) tensile test to failure conducted by Carter et al.[47,48] resulted in widely varying yield strengths. The yield strains of all specimens, however, exhibited far less data scatter. This finding is consistent with the view of Currey,[11] who suggested that bone yielding is more strongly controlled by strain magnitude than by stress magnitude. The fatigue behavior of the bone specimens was also found to be much more dependent on strain range than on stress range. The plot of strain range vs. cycles to failure exhibited much less data scatter than the plot of initial stress range versus cycles to failure. Further statistical analysis of the elastic modulus of the specimens tested showed that a great deal of the scatter observed in the stress range fatigue curve could be explained on the basis of specimen modulus. Variations in modulus, however, had very little influence on the fatigue behavior of cortical bone as shown in the plot of strain range vs. cycles to failure.

VI. ACOUSTIC EMISSION STUDIES

Acoustic emission may be defined as the elastic stress waves generated by the rapid release of energy during dynamic processes in a material. Seismic shocks, for example, are acoustic emission on a very large scale caused by the release of energy during slip along a fault in the earth's crust. In general, there are two types of acoustic emission. Continuous emission is composed of a multitude of small-amplitude events (or energy releases) superimposed on one another. Burst emission, on the other hand, consists of distinct high-amplitude events. Much of the work performed to detect and study material damage using acoustic emission has concentrated on detecting and discriminating these two types of emission. The application of acoustic emission to composite materials and to the problem of interfacial debonding is of special importance to the study of bone. Henneke et al.[49] and Fitz-Randolph et al.[50] studied acoustic emission from mechanical tests on a boron-epoxy composite and were able to correlate acoustic emission with the yield load of the material and with fiber breakage occurring during crack propagation. Other studies have made similar findings in other composite systems.[51-53]

Netz et al.[54] used acoustic emission techniques to study the torsional loading characteristics of whole canine femurs and tibias. The bones were loaded at a twist rate of 6°/sec until complete bone fracture resulted. The torque-displacement curves were recorded simultaneously with the acoustic emission counts. The torque-displacement curve for all bones was initially linear. The proportional limit, as determined grossly from the torque angle recordings, occurred within the range of 30 to 50% of the total twist angle at failure. Acoustic emission signals were recorded only after the proportional limit was reached. The signals, characteristic of crack formation and/or growth, appeared in spurts with increasing frequency as the ultimate failure of the bone was approached (Figure 9). Netz and co-workers felt that their study indicated that the nonlinear deformation of cortical bone prior to ultimate failure is due to the gradual formation of microcracks. The entire diaphysis at all torque loads behaves, they argued, as a structure composed of an "elastic-cracking" material.

Wright et al.[55] recorded acoustic emission signals during monotonic tension tests from standardized specimens machined from bovine femurs. The tensile stress-strain curves were recorded simultaneously with the acoustic emission counts. Testing was done with fully wet specimens at a strain rate of approximately 0.025/sec. A clip-on strain gage extensometer was used to measure specimen elongation. The recorded stress-strain curves were similar to those reported in previous investigations (Figure 10). Acoustic emission signals began to appear as the strain approached 60 to 70% of the macroscopic yield strain. As gross yielding occurred, there was an accumulation of acoustic emission signals. Few emissions occurred in the major portion of the postyield regime. Immediately prior to ultimate failure, there was an abrupt accumulation of high-amplitude signals.

VII. MICROSCOPIC EXAMINATION OF YIELD AND FATIGUE DAMAGE

Sweeney et al.[56] demonstrated that the fracture surfaces of longitudinal compact bone specimens loaded to failure were approximately perpendicular to the direction of applied loading. Compression loading of similar specimens, however, created oblique fracture surfaces corresponding to planes of high shear stress. Similar results in axial loading were obtained by Reilly.[57] This observation suggests that the micromechanical mechanisms associated with yield are quite different in tensile and compressive loading.

To investigate the failure mechanisms in cortical bone, Currey and Brear[58] conducted a microscopic examination of bone yielding. Standard test specimens were machined and their surfaces were polished with a fine alumina paste. The specimens were loaded in either bending or compression. Some of the specimens were loaded only in the elastic region of the load-displacement curve, others were loaded into the postyield regions, and some were loaded until complete failure occurred. After testing, the specimens were placed in a stain for 1 to 6 hr at room temperature. The stain was 0.5 g basic fuchsin, 1 g xylidine ponceau, 20 mℓ absolute ethanol, and 0.05% acetic acid to 200 mℓ. The specimens were washed in water and inspected microscopically.

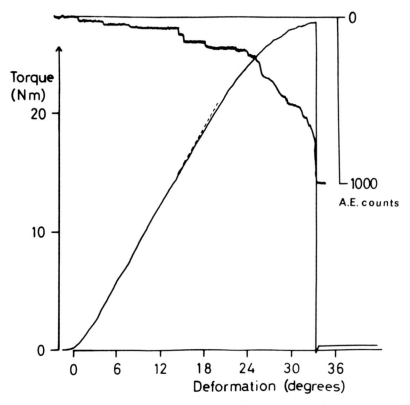

FIGURE 9.   Torque-displacement curve with simultaneously recorded cumulative acoustic emission counts from a canine long bone. (From Netz, P., Eriksson, K., and Stromberg, L., *Acta Orthop. Scand.*, 51, 223, 1980. With permission.)

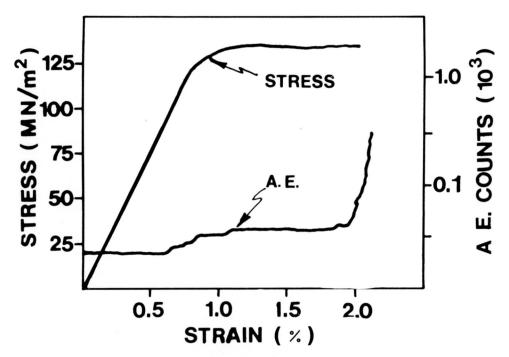

FIGURE 10.   Stress-strain curve with acoustic emission counts for cortical bone specimen tested in tension. (From Wright, T. M., Vosburgh, F., and Burstein, A. H., *J. Biomech.*, 14, 405, 1981. With permission.)

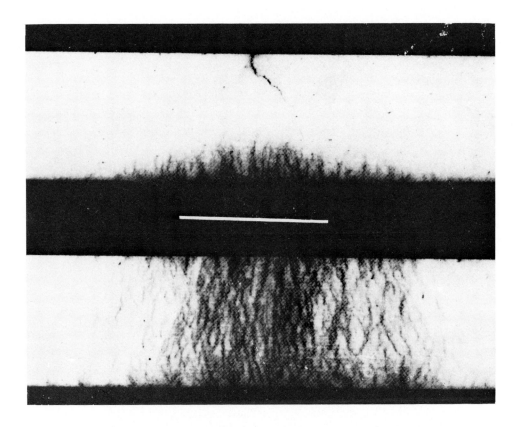

FIGURE 11.   View of the side (above) and of the tension surface (below) of a bone specimen loaded in three-point bending. Yield damage is accentuated by a strain. (From Currey, J. D. and Brear, K., *Calcif. Tissue Res.*, 15, 173, 1974. With permission.)

In the bending specimens loaded only in the elastic region, no clear staining was visible except in blood vessels. In specimens that had yielded, however, numerous diffuse fine lines were seen which passed from the tensile surface toward the interior of the specimens (Figure 11). Sometimes there were also shear lines (oriented at oblique angles to the loading direction) on the compression side of the specimens, but these were distinctly different in appearance from the tensile lines (Figure 12).

In a study similar to that of Currey and Brear, Carter and Hayes[59] conducted a microscopic investigation of the damage created in cortical bone specimens exposed to cyclic fatigue bending loads. In addition, the fracture surfaces of specimens fatigued to failure were examined by electron microscopy and compared to those created in monotonic bending tests to failure. The fracture pattern of specimens subjected to flexural fatigue was similar to that of specimens fractured by a single applied loading. A transverse fracture surface was created on the tension side of each of the specimens, and an oblique fracture surface was formed on the compression side (Figure 13). In all cases, however, the oblique fracture surface of each of the fatigue specimens was larger than that observed for the monotonic bending specimens.

Specimens that were not fatigued to complete fracture exhibited diffuse microscopic damage. To illustrate this general finding, the loading history and photomicrographs taken from a representative specimen of primary bovine bone subjected to six loading cycles were presented. The load amplitudes of each cycle for this specimen were such that noticeable yielding occurred prior to unloading. Repeated loading caused a progressive loss of stiffness,

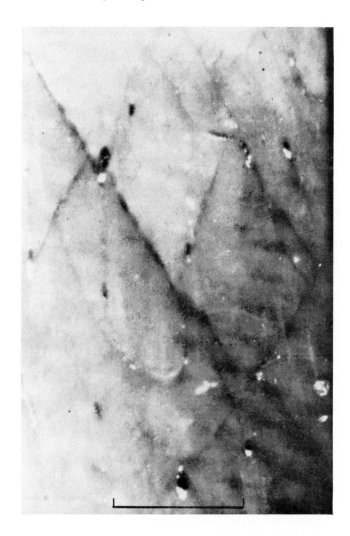

FIGURE 12.   Photograph of bone loaded in compression (from top to bottom) showing cracks and oblique shear lines; marker, 200 μm. (From Currey, J. D. and Brear, K., *Calcif. Tissue Res.*, 15, 173, 1974. With permission.)

a decrease in yield strength, and an increase in permanent deformation and hysteresis. This loading behavior is indicative of diffuse structural damage. Loading was terminated after the sixth cycle to prevent complete specimen fracture.

Examination of the specimen surface after testing revealed that damage was most extensive on the compression side. The major damage modes on the compression side were oblique cracking and longitudinal splitting (Figure 14A). Microcracking was extensive throughout the region and appeared to be influenced by stress concentrations created by vascular canals and lacunae (Figure 14B). Longitudinal splitting occurred through weak interfaces created by layers of primary osteons and interlamellar cement bands. Examinations at higher magnification revealed microcracks through or near the lacunae (Figure 14C).

The damage observed on the tension side of the specimen was more subtle and consisted primarily of separation (or debonding) at cement lines and interlamellar cement bands (Figure 15A and B). Occasional microcracking of interstitial bone was also observed. High magnifications of debonding around an osteon revealed significant fibrous tearing (Figure 15C).

Lakes and Saha[60] examined the creep behavior of cortical bone specimens subjected to

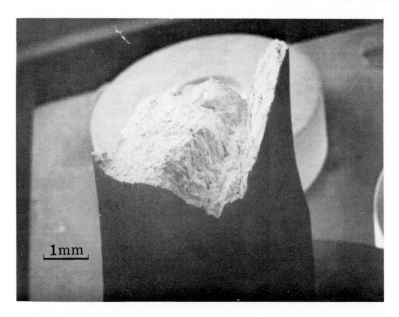

FIGURE 13.   Fracture surface of bone specimen tested in one directional flexural fatigue. The tension side is to the left where a transverse fracture surface was created. The compression side is to the right where an oblique fracture surface was created on a plane of high shear stress. (From Carter, D. R. and Hayes, W. C., *Clin. Orthop.*, 127, 265, 1977. With permission.)

constant torsional loading for up to 42 days. When high torsional loads were applied, the torsional creep was quite striking. The specimen compliance over the 42-day testing period increased by more than a factor of 4. After unloading, a significant residual strain remained in the specimens. To identify the microstructural origin of this permanent strain, extremely fine fiduciary lines were scribed on the bone surface prior to testing. After testing these lines were examined microscopically. The displacements of the fiduciary marks indicated shear slippage at the cement band interfaces of the bone lamellae (Figure 16). Displacements of this type were not observed in specimens which were fractured rapidly in torsion. Lakes and Saha concluded that the process of viscous-like slippage at bone cement substance interfaces is a dominant mechanism responsible for the large creep deformations and permanent strain observed.

In their microstructural testing investigations of single osteons, Ascenzi and Bonucci conducted microscopic examinations of the damage created. For the specimens tested in compression[35] an extensive network of fractures was created in the osteons (Figure 17). These defects varied in number within the same osteon sample and among the different osteons tested. Fractures and fissures crossed at oblique angles to the osteon axis at angles of between 30 to 35°. The pattern of the fissures was apparently independent of the osteon microstructure and degree of calcification. Photomicrographs taken at the magnification of 1400 × showed that the fissures continued uninterrupted through the lamellae.

Ascenzi and Bonucci[36] studied microscopically the tensile damage in osteons with alternating lamellae in an effort to explain the mechanical behavior depicted in Figure 6. Their electron photomicrographs revealed that these osteons underwent morphological alterations before the ultimate tensile strength was achieved. Many fracture-like spaces were evident in the lamellae with fibers oriented transversely to the osteon axis. Generally, these spaces crossed the lamellae in a direction perpendicular to the osteon axis (the direction of applied loading), although some irregularity was noted. When the fracture-like spaces were wide, they extended across the entire lamellae until the boundaries of two lamellae with longitu-

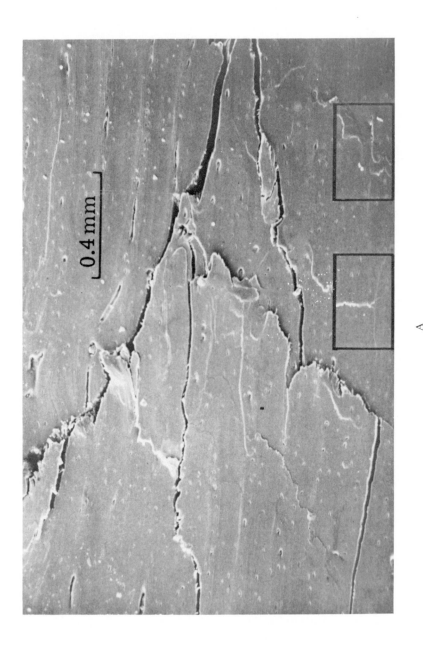

FIGURE 14.   Compression fatigue damage. (A) SEM photomicrograph of compression side of flexural fatigue specimen. (B) Compression microcracks in an area shown in (A). (C) Microcrack near a lacuna shown in (B). (From Carter, D. R. and Hayes, W. C., *Clin. Orthop.*, 127, 265, 1977. With permission.)

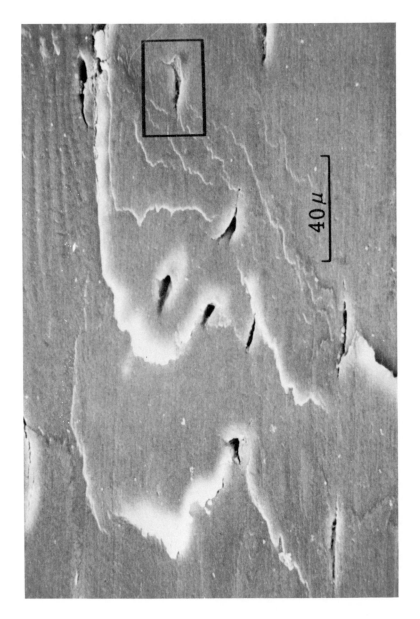

Figure 14B.

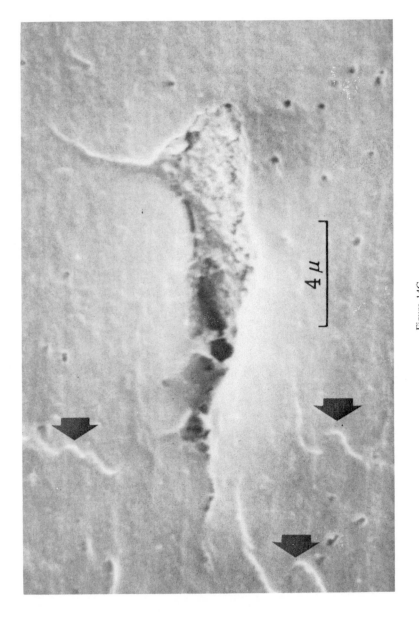

Figure 14C.

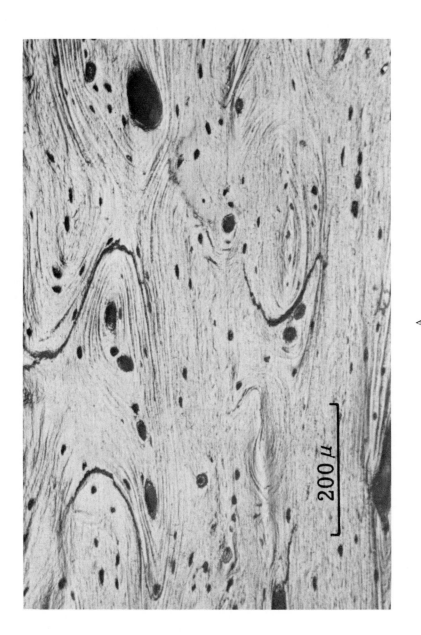

A

FIGURE 15. Reflected-light photomicrograph of tension side of flexural fatigue specimen; (B) SEM photomicrograph depicting osteon debonding shown in (A); (C) higher magnification of osteon debonding area shown in (B). (From Carter, D. R. and Hayes, W. C., *Clin. Orthop.*, 127, 265, 1977. With permission.)

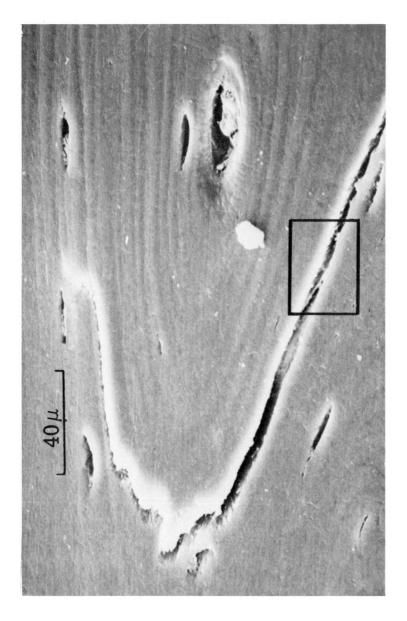

Figure 15B.

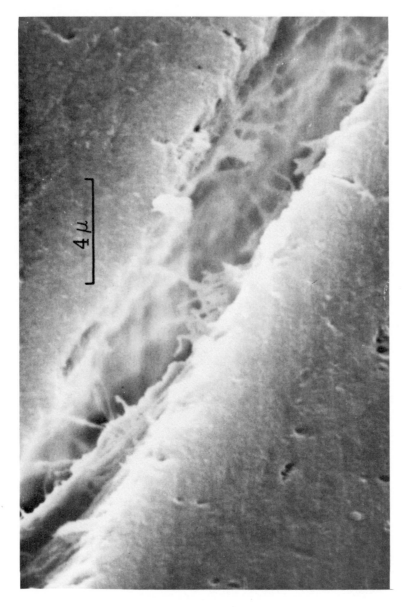

Figure 15C.

FIGURE 16.   Micrograph of the surface of bovine bone specimen tested in a torsional creep mode. Fiduciary marks were straight and unbroken prior to loading and after loading they indicate residual strain facilitated by deformation in the interlamellar cement bands; marker, 100 μm. (From Lakes, R. and Saha, S., *Science*, 204, 501, 1979. With permission.)

dinally oriented fibrils were reached (Figure 18A and B). Using ultrathin sections from decalcified osteons, they observed that some of the fracture-like spaces were derived from true discontinuities at the boundaries between collagen fibrils (i.e., at the level of the interfibrillar cementing substance). Other spaces were interpreted as empty osteocyte canaliculi that were probably enlarged by loading.

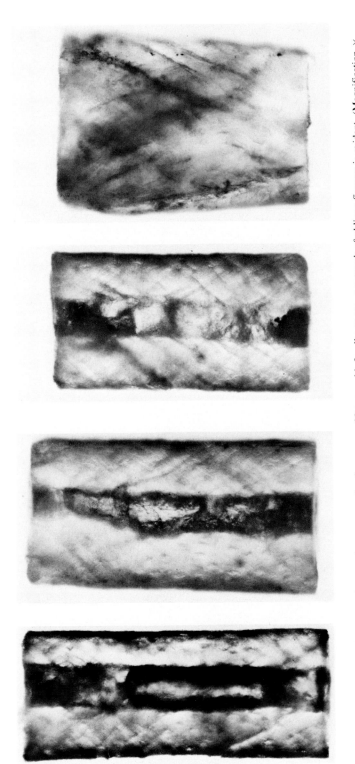

FIGURE 17. A series of compression tested osteon samples from a man 30 years old. In all osteons, a network of oblique fissures is evident. (Magnification × 100.) (From Ascenzi, A. and Bonucci, E., *Anat. Rec.*, 161, 377, 1968. With permission.)

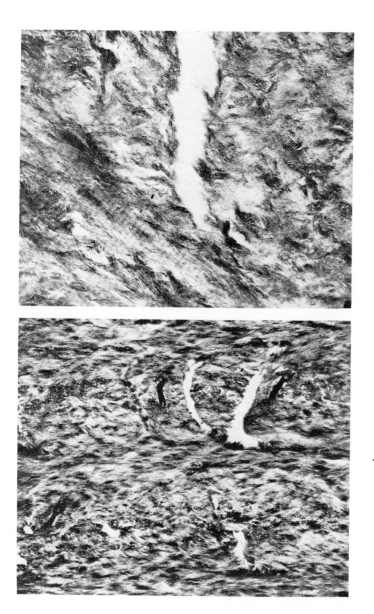

B

A

FIGURE 18.   Bone tissue damage caused by tensile loading. (A) Electron microphotograph of an alternate osteon loaded under tension. Lamellae with fiber bundles having transversal course are wider than those with longitudinally oriented fiber bundles and reveal many fracture-like interruptions. (Undecalcified and unstained; magnification × 13,000.) (B) Electron microphotograph of a fracture-like interruption in a lamella with fiber bundles having transversal course.   The edges of the interruption are irregular and appear torn. (Undecalcified and unstained; magnification × 52,000.) (From Ascenzi, A. and Bonucci, E., *J. Biomech.*, 9, 65, 1976. With permission.)

REFERENCES

1. **Reilly, D. T., Burstein, A. H., and Frankel, V. H.,** The elastic modulus for bone, *J. Biomech., 7*, 271, 1974.
2. **Reilly, D. T. and Burstein, A. H.,** The elastic and ultimate properties of compact bone tissue, *J. Biomech., 8*, 393, 1975.
3. **McElhaney, J. H.,** Dynamic response of bone and muscle tissue, *J. Appl. Physiol., 21*, 1231, 1966.
4. **Frankel, V. H. and Burstein, A. H.,** *Orthopaedic Biomechanics,* Lea & Febiger, Philadelphia, 1970.
5. **Wright, T. M. and Hayes, W. C.,** Tensile testing of bone over a wide range of strain rates: effects of strain rate, microstructure and density, *Med. Biol. Eng., 14*, 671, 1976.
6. **Evans, F. G. and Lebow, M.,** Regional differences in some physical properties of the human femur, *J. Appl. Physiol., 3*, 563, 1951.
7. **Burstein, A. H., Currey, J. D., Frankel, V. H., and Reilly, D. T.,** The ultimate properties of bone tissue: the effects of yielding, *J. Biomech., 5*, 34, 1972.
8. **Evans, F. G.,** *Mechanical Properties of Bone,* Charles C Thomas, Springfield, Ill., 1973.
9. **Hayes, W. C. and Carter, D. R.,** Biomechanics of bone, in *Skeletal Research — An Experimental Approach,* Simmons, D. J. and Kunin, A., Eds., Academic Press, New York, in press.
10. **Burstein, A. H., Zika, J. M., Heiple, K. G., and Klein, L.,** Contribution of collagen and mineral to the elastic-plastic properties of bone, *J. Bone Jt. Surg., 57a*, 956, 1975.
11. **Currey, J. D.,** The mechanical properties of bone, *Clin. Orthop., 73*, 210, 1970.
12. **Bell, G. H., Khogali, A., and Sharma, D. N.,** Quantitative studies on the suppression of skeletal lesions in the rat produced by β-aminoproprionitrile, *Q. J. Exp. Physiol., 47*, 244, 1962.
13. **Bornstein, P.,** The cross-linking of collagen and elastin and its inhibition in osteolathyrism. Is there a relation to the aging process?, *Am. J. Med., 49*, 429, 1970.
14. **Henneman, D. H.,** Inhibition by estradiol-17 β of the lathyritic effect of β-aminoproprionitrile on skin and bone collagen, *Clin. Orthop., 83*, 245, 1972.
15. **Spengler, D. M., Baylink, D. J., and Rosenquist, J. B.,** Effects of β-aminoproprionitrile on bone mechanical properties, *J. Bone Jt. Surg., 59a*, 670, 1977.
16. **Rosenquist, J. B., Baylink, D. J., and Spengler, D. M.,** The effect of BAPN on bone mineralization, *Proc. Soc. Exp. Biol. Med., 154*, 310, 1977.
17. **Weir, J. B de V., Bell, G. H., and Chambers, J. W.,** The strength and elasticity of bone in rats on a rachitogenic diet, *J. Bone Jt. Surg., 31b*, 444, 1949.
18. **Rasmussen, D. M., Wakin, K. G., and Winkelman, R. K.,** Isotonic and isomeric thermal contraction of human dermis. II. Age-related changes, *J. Invest. Dermatol., 43*, 341, 1964.
19. **Bailey, A. J.,** The nature of collagen, in *Comprehensive Biochemistry,* Florkin, M. and Stotz, E. H., Eds., Vol. 26B, Elsevier, Amsterdam, 1968, 297.
20. **Verzar, F.,** The aging of collagen, *Int. Rev. Connect. Tissue Res., 2*, 243, 1964.
21. **Fujii, K., Yuboki, Y., and Sasaki, S.,** Aging of human bone and articular cartilage collagen, *Gerontology, 22*, 363, 1976.
22. **Currey, J. D. and Butler, G.,** The mechanical properties of bone tissue in children, *J. Bone Jt. Surg., 57a*, 810, 1975.
23. **Black, J., Mattson, R., and Korostoff, E.,** Haversian osteons: size, distribution, internal structure, and orientation, *J. Biomed. Material Res., 8*, 299, 1974.
24. **Kerley, E. R.,** The microscopic determinants of age in human bone, *Am. J. Phys. Anthropol., 23*, 149, 1965.
25. **Vinz, H.,** Die anderung der festigkeitseigenschaften des kompakten knochengewebes im laufe der altersentwicklung, *Gegenbaurs Morphol. Jahrb., 115*, 257, 1970.
26. **Vinz, H.,** Die festigkeit der renen knochensubstanz. Naherungsverfahren zur bestimmung der auf den hohlraumfrein querschnitt bezogenen festigkeit von knochengewebe, *Gegenbaurs Morphol. Jahrb., 117*, 453, 1972.
27. **Burstein, A. H., Reilly, D. T., and Martens, M.,** Aging of bone tissue: mechanical properties, *J. Bone Jt. Surg., 58a*, 82, 1976.
28. **Bonfield, W. and Li, C. H.,** Deformation and fracture of bone, *J. Appl. Phys., 37*, 869, 1966.
29. **Carter, D. R.,** Anisotropic analysis of strain rosette information from cortical bone, *J. Biomech., 11*, 199, 1978.
30. **Carter, D. R., Smith, D. J., Spengler, D. M., Daly, C. H., and Frankel, V. H.,** Measurement and analysis of *in vivo* bone strains on the canine radius and ulna, *J. Biomech., 13*, 27, 1980.
31. **Black, J.,** Dead or alive: the problem of *in vitro* tissue mechanics, *J. Biomed. Material Res., 10*, 377, 1976.
32. **Bonfield, W. and O'Connor, P.,** Anelastic deformation and the friction stress of bone, *J. Material Sci., 13*, 202, 1978.
33. **Bonfield, W. and Li, C. H.,** Anisotropy of nonelastic flow in bone, *J. Appl. Phys., 38*, 2450, 1967.

34. **Ascenzi, A. and Bonucci, F.,** The tensile properties of single osteons, *Anat. Rec.,* 158, 375, 1967.
35. **Ascenzi, A. and Bonucci, E.,** The compressive properties of single osteons, *Anat. Rec.,* 161, 377, 1968.
36. **Ascenzi, A. and Bonucci, E.,** Mechanical similarities between alternate osteons and cross-ply laminates, *J. Biomech.,* 9, 65, 1976.
37. **Reed-Hill, R. E.,** *Physical Metallurgy Principles,* Van Nostrand, Princeton, N.J., 1964.
38. **Salkind, M. J.,** Fatigue of composites, in *Composite Materials: Testing and Design,* American Society for Testing and Materials, Philadelphia, 1972, 143.
39. **Yang, J. N. and Trapp, W. J.,** Reliability analysis of aircraft structures under random loading and periodic inspection, *AIAA J.,* 12, 1623, 1974.
40. **Allen, H. G.,** Stiffness and strength of two glass-fiber reinforced cement laminates, *J. Compos. Material,* 5, 194, 1971.
41. **Owen, M. J. and Howe, R. J.,** The accumulation of damage in a glass-reinforced plastic under tensile and fatigue loading, *J. Phys. D.,* 5, 1637, 1972.
42. **Burstein, A. H., Reilly, D. T., and Frankel, V. H.,** Failure characteristics of bone and bone tissue, in *Perspectives in Biomedical Engineering,* Kenedi, R. M., Ed., University Park Press, London, 1973, 131.
43. **Currey, J. D.,** The effects of strain rate, reconstruction, and mineral content on some mechanical properties of bovine bone, *J. Biomech.,* 8, 81, 1975.
44. **Carter, D. R. and Hayes, W. C.,** Compact bone fatigue damage. I. Residual strength and stiffness, *J. Biomech.,* 10, 325, 1977.
45. **Neville, A. M.,** *Properties of Concrete,* Pitman, New York, 1973.
46. **Reilly, D. T. and Burstein, A. H.,** The mechanical properties of cortical bone, *J. Bone Jt. Surg.,* 56a, 1001, 1974.
47. **Carter, D. R., Caler, W. E., Spengler, D. M., and Frankel, V. H.,** Fatigue behavior of adult cortical bone — the influence of mean strain and strain range, *Acta Orthop. Scand.,* in press.
48. **Carter, D. R., Caler, W. E., Spengler, D. M., and Frankel, V. H.,** Uniaxial fatigue of human cortical bone. The influence of tissue physical characteristics, *J. Biomech.,* 14, 461, 1981.
49. **Henneke, E. G., II, Herakovich, C. T., Jones, G. L., and Renieri, M. P.,** Acoustic emission from composite - reinforced metals, *Exp. Mech.,* 15, 10, 1975.
50. **Fitz-Randolph, J. M., Phillips, D. C., Beaumont, P. W. R., and Tetelman, A. S.,** The fracture energy and acoustic emission of a boron-epoxy composite, *J. Material Sci.,* 7, 289, 1972.
51. **Mehan, R. L. and Mullin, J. V.,** Analysis of composite failure mechanisms using acoustic emissions, *J. Compos. Material,* 5, 266, 1971.
52. **Chang, R. H., Gordon, D. E., and Gardner, A. H.,** A study of fatigue damage in composites by nondestructive testing techniques, in *Fatigue of Filamentary Composite Materials,* American Society for Testing and Materials, Philadelphia, 1977, 57.
53. **Carlyle, J. M.,** Imminent fracture detection of graphite/epoxy using acoustic emission, *Exp. Mech.,* 18, 191, 1978.
54. **Netz, P., Eriksson, K., and Stromberg, L.,** Material reaction of diaphyseal bone under torsion — an experimental study on dogs, *Acta Orthop. Scand.,* 51, 223, 1980.
55. **Wright, T. M., Vosburgh, F., and Burstein, A. H.,** Permanent deformation of compact bone monitored by acoustic emission, *J. Biomech.,* 14, 405, 1981.
56. **Sweeney, A. W., Byers, R. K., and Kroon, R. P.,** Mechanical characteristics of bone and its constituents, ASME paper 65-WA/HUF-7, 1965.
57. **Reilly, D. T.,** The Mechanical Properties of Bone, Ph.D. dissertation, Case Western Reserve University, Cleveland.
58. **Currey, J. D. and Brear, K.,** Tensile yield in bone, *Calcif. Tissue Res.,* 15, 173, 1974.
59. **Carter, D. R. and Hayes, W. C.,** Compact bone fatigue damage: a microscopic examination, *Clin. Orthop.,* 127, 265, 1977.
60. **Lakes, R. and Saha, S.,** Cement line motion in bone, *Science,* 204, 501, 1979.
61. **Carter, D. R. and Spengler, D. M.,** Mechanical Properties and chemical composition of cortical bone, *Clin. Orthop.,* 135, 192, 1978.
62. **Bonfield, W.,** personal communication.

Chapter 3

MACROSCOPIC DIRECTIONALITY IN BONE

Timothy M. Wright and Dennis R. Carter

TABLE OF CONTENTS

I.   Introduction 38

II.   Microstructure and Elastic Symmetry 38

III.   In Vivo Stresses and Strains 41

IV.   Multidirectional Failure Theories 44

V.   Concluding Remarks 46

References 48

I. INTRODUCTION

To describe the mechanical performance of bone and its directionality dependence requires an understanding of bone's composite structure at the ultrastructural level, the microstructural level, and at the level of the organ itself. Ultrastructurally, the mechanical properties of bone tissue are dominated by the contributions of collagen fibers and hydroxyapatite crystals and the elusive bond between them. Unlike common engineering composites, bone has its fibrous structure (collagen) as the matrix rather than as the strong, reinforcing element of the composite. Microstructurally, bone again exhibits a composite nature, consisting of lamina of tissue arranged in an organized pattern and separated by comparably weak interfaces. The pattern of organization can vary from species to species and with age or disease within a species. The most common in adult human bone is, of course, the osteon. Even at the level of the organ, bone can be considered as a composite of sorts, consisting of dense cortical bone to form the shafts and peripheries and trabecular bone to provide an internal means of distributing applied loads.

The tubular shape of whole long bones is effective in resisting both applied torques and bending moments. These types of loading are common in vivo. The tubular shape optimizes the structure's resistance to failure under these loads while minimizing the amount of material necessary. This same optimization scheme has long been noted in the trabecular architecture of cancellous bone (in the proximal femur, for example).

In this chapter, the properties which describe the mechanical behavior of cortical bone will be reviewed. Emphasis will be placed on the influence that microstructural organization has on these properties.

II. MICROSTRUCTURE AND ELASTIC SYMMETRY

The microstructure of cortical bone is most easily understood on a developmental basis. In this manner, bone can be either primary or secondary. Primary bone is the first bone formed in a region, either by endochondral ossification or subperiosteal deposition. In humans, primary bone consists of circumferential bone (with orderly collagen bundles) and primary osteons (made of successive lamellae of bone with one or more vascular canals).[1,2] Primary osteons are generally oriented parallel to the long bone axis and are surrounded by woven-fibered bone, a third type of primary bone with randomly oriented collagen bundles.[2,3]

With aging, the continuous process of remodeling significantly alters the microstructure. Osteoclastic activity results in bone resorption. The resorption occurs in a direction generally parallel to the long bone axis. The anastomosing tubular cavities are filled by osteoblastic activity with successive concentric lamellae of bone until a single vascular canal remains at the center. The resulting cylinders are called secondary osteons.[4,5] Secondary osteons are always surrounded by cement lines marking the edge of the osteoclastic activity and the starting point of the osteoblastic bone formation. The development of cortical bone microstructure is reviewed in Figure 1 and discussed in more detail by Carter et al.[1]

The longitudinal orientation of secondary osteons led many researchers to suggest that bone was anisotropic (exhibiting different elastic properties in different directions). Dempster and Liddicoat,[6] for example, performed compression tests on small cubes of human cortical bone loaded in directions corresponding to the longitudinal, transverse, and radial axes of the bone (Figure 2). The results showed that the elastic modulus in the longitudinal direction was approximately twice as large as the nearly equal moduli measured in the other two directions. Though the specimen preparation technique and specimen geometry can be criticized,[7] the work stands as one of the earliest attempts to examine anisotropy in bone.

More recently, two techniques were used to determine the degree of anisotropy in compact bone. Lang[8] performed ultrasonic measurements on dry and fresh bovine bone. A transversely

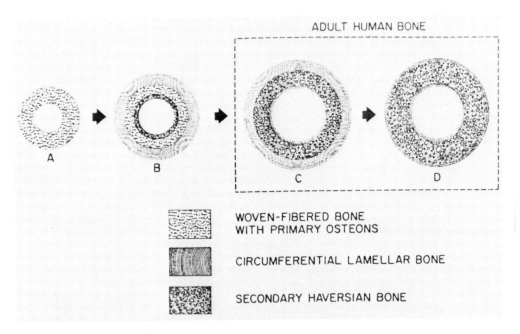

FIGURE 1.   Maturation of human cortical bone. (From Carter, D. R., Hayes, W. C., and Schurman, D. J., *J. Biomech.*, 9, 213, 1976. With permission.)

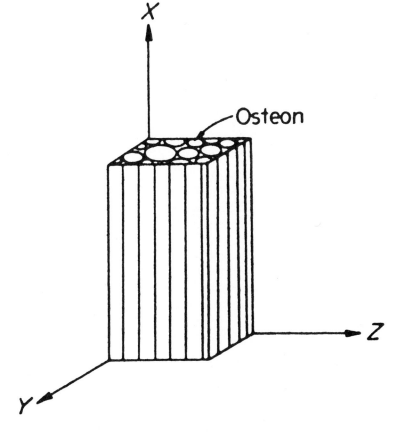

FIGURE 2.   Orientation of axes with respect to bone structure. (From Reilly, D. T. and Burstein, A. H., *J. Biomech.*, 8, 394 and 402, 1975. With permission.)

Table 1
STRESS-STRAIN RELATIONSHIP FOR A TRANSVERSELY ISOTROPIC MATERIAL SUCH AS BONE[a]

$$
\begin{Bmatrix} \sigma_1 \\ \sigma_2 \\ \sigma_3 \\ \tau_{23} \\ \tau_{31} \\ \tau_{12} \end{Bmatrix}
=
\begin{bmatrix}
C_{11} & C_{12} & C_{13} & 0 & 0 & 0 \\
C_{12} & C_{11} & C_{13} & 0 & 0 & 0 \\
C_{13} & C_{13} & C_{33} & 0 & 0 & 0 \\
0 & 0 & 0 & C_{44} & 0 & 0 \\
0 & 0 & 0 & 0 & C_{44} & 0 \\
0 & 0 & 0 & 0 & 0 & (C_{11} - C_{12})/2
\end{bmatrix}
\begin{Bmatrix} \epsilon_1 \\ \epsilon_2 \\ \epsilon_3 \\ \gamma_{23} \\ \gamma_{31} \\ \gamma_{12} \end{Bmatrix}
$$

[a]   With σ = normal stress, τ = shear stress, ϵ = normal strain, γ = shear strain, C = elastic stiffness coefficients, and the three-direction being longitudinal for the case of cortical bone.

Table 2
INVESTIGATIONS OF THE ELASTIC PROPERTIES OF COMPACT BONE

Investigator	Material	Stiffness coefficients[a]					Engineering constants[b]				
		C_{11}	C_{33}	C_{12}	C_{13}	C_{44}	E_1	E_3	ν_{12}	ν_{31}	G_{13}
Lang (1969)	Dry bovine femur	2.38	3.34	1.02	1.12	0.82	1.80	2.60	0.32	0.33	0.82
	Fresh bovine phalanx	1.97	3.20	1.21	1.26	0.54	1.13	2.20	0.48	0.40	0.54
Reilly and Burstein (1975)	Fresh bovine femur	1.70	2.97	1.03	0.98	0.36	1.02	2.26	0.51	0.36	0.36
	Fresh human femur	4.66	5.34	3.94	3.96	0.33	1.15	1.70	0.58	0.46	0.33
Yoon and Katz (1976)	Dry human femur	2.34	3.25	0.91	0.91	0.87	1.88	2.74	0.31	0.28	0.87

[a]   Stiffness coefficients are in units of 10^{10} N/m^2.

[b]   Elastic (E) and shear (G) moduli are in units of 10^{10} N/m^2. Poisson's ratios are dimensionless.

isotropic model was assumed such that one set of elastic properties would be exhibited in the longitudinal direction and a second set in the plane perpendicular to that direction (i.e., the plane containing the transverse and radial directions). Such a material has a constitutive relation between the components of stress and strain that contains five independent constants (the stiffness coefficients) as shown in Table 1.[9] Lang calculated these constants from the measured ultrasonic wave speeds in various directions through the bone microstructure and the measured density of the bone specimens used. With the same technique, Yoon and Katz[10] performed a similar experiment on dry human cortical bone. The resulting stiffness coefficients from the two studies are summarized in Table 2 along with the corresponding engineering constants.

The second technique used to determine the degree of anisotropy involved quasistatic tensile testing of carefully prepared standardized specimens of fresh human and bovine bone. Reilly and Burstein[11] performed uniaxial tests on specimens taken in the longitudinal, circumferential, and radial directions (as well as in directions 30 and 60° to the long axis). By using clip-on extensometers, strains were measured both parallel and perpendicular to the direction of loading. This allowed both an elastic modulus, E, and a Poisson's ratio, ν, to be measured for each specimen. Poisson's ratio is the strain measured perpendicular to the loading direction to the strain measured parallel to the loading direction. The transverse modulus from the circumferential and radial specimens was found to be 11.5 GN/m^2. The longitudinal modulus was 17.0 GN/m^2, approximately 50% greater than the transverse modulus. Combined with shear modulus, G, which was measured from torsion tests on similar longitudinal specimens, all five independent engineering constants for the transversely isotropic model were obtained (see Table 2).

Results from studies such as those described here are important for several reasons. They reinforce current understanding of the symmetry in bone microstructure by demonstrating a model for elastic behavior. Also, they provide the information necessary to interpret in vivo bone strain measurements. Finally, they point out the need for a multidirectional failure theory for bone; if the elastic behavior is anisotropic, then any failure strength theory must encompass anisotropy. The remainder of this chapter will deal with these last two points.

III. IN VIVO STRESSES AND STRAINS

Any consideration of the mechanical behavior of bone requires knowledge of the applied stresses the bone material must withstand in vivo. In vitro whole bone experiments which attempt to recreate in vivo loading conditions are beneficial in this regard, as are analytical techniques. However, both methods suffer to some extent from necessary oversimplification of the complex geometry and numerous applied loads existing in vivo.

To overcome these difficulties, several investigators have employed strain gauge techniques to perform experimental stress analysis of bone.[12-17] Lanyon[14] first bonded rosette strain gauges to a sheep calcaneus in vivo, enabling him to calculate the magnitudes and directions of principal strains. Later, Lanyon and co-workers[15] applied the same technique in collecting strain data from the anteromedial aspect of a human tibia during walking and jogging. The resulting strains from the three gauges (Figure 3) were again used to calculate the two principal strains and their orientation with respect to the long axis of the tibia.

Recently, Carter applied the transversely isotropic model for cortical bone to this data to convert the measured strains to stresses. In an isotropic material, such an analysis is straightforward. The orientation of the principal stresses is identical to that of the principal strains and the magnitudes of the principal stresses are easily calculated from the two isotropic elastic constants and the principal strains.

Stress analyses based on strain readings from an anisotropic material like bone are more complicated. The orientation of the principal stresses, for example, is not coincident with that of the principal strains. Also, two material constants are insufficient to perform the analysis. For bone, four material constants are required for the two-dimensional plane stress case. Furthermore, the constitutive relations between stress and strain are dependent upon orientation with respect to the principal material directions:

$$\begin{bmatrix} \sigma_1 \\ \sigma_3 \\ \tau_{13} \end{bmatrix} = \underline{C} \begin{bmatrix} \epsilon_1 \\ \epsilon_3 \\ \gamma_{13} \end{bmatrix} \qquad (1)$$

where σ_1, σ_3, τ_{13} = stresses in the material directions and ϵ_1, ϵ_3, γ_{13} = strains in the material directions. For bone, the principal material directions are generally assumed parallel to the long axis of the bone (three-axis) and perpendicular to the long axis (one-axis). $\underline{C}$ in Equation 1 is the stiffness matrix (see Table 1) reduced to the plane stress case ($\sigma_2 = \tau_{12} = \tau_{22} = 0$):

$$\underline{C} = \begin{bmatrix} C_{11} & C_{13} & 0 \\ C_{13} & C_{33} & 0 \\ 0 & 0 & C_{44} \end{bmatrix} \qquad (2)$$

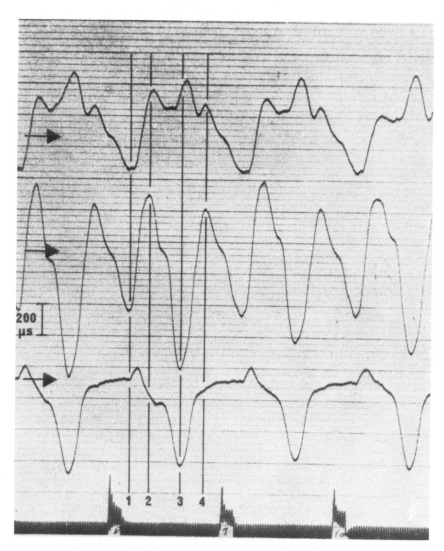

FIGURE 3.   Strain gauge data during walking (from three gauges attached to the human tibia in vivo). Arrows mark the level of zero strain. Points 1 through 4 denote heel strike, full foot contact with heel off, toe off, and forward swing, respectively. (From Lanyon, L. E., Hampson, W. G. J., Goodship, A. G., and Shan, J. S., *Acta Orthop. Scand.*, 46, 257, 1975. With permission.)

Carter used the strain data of Lanyon transformed to the material directions and the stiffness matrix (Equation 2) obtained from Reilly and Burstein's experimental data (see Table 2) to solve Equation 1. The calculated stresses in the principal material directions are shown in Figure 4 for one gait cycle. The results showed that during normal walking the stress along the tibial shaft was compressive during heel strike and push-off, but tensile during the stance phase. The stress perpendicular to the long axis of the tibia was negligible, but a significant shear stress was present near the end of the stance phase and during push-off (indicating torsional loading in external rotation). During jogging, the stress along the axis of the tibia was again compressive when the foot contacted the ground, but at push-off a large tensile stress developed longitudinally.

The stress analysis described in Equations 1 and 2 can also be used to calculate principal stresses and their orientation, though this requires a transformation of the stiffness matrix.[18] Performing this analysis for the normal walking data of Lanyon, Carter found that the

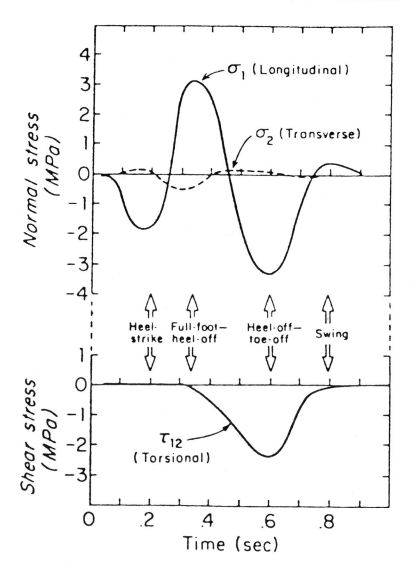

FIGURE 4. The calculated stresses in the principal material directions calculated from the strain data for normal walking in Figure 3. Note that in the figure the 1 and 2 directions correspond to the text to the 3 and 1 directions, respectively. (From Carter, D. R., *J. Biomech.*, 11, 201, 1978. With permission.)

principal stresses were oriented at an angle of $-58°$ to the longitudinal axis (i.e., 58° in an anterolateral direction from the proximal direction along the long axis). This differs from the principal strain orientation, which was at an angle of $-53°$. The findings of Lanyon and Carter are summarized in Table 3 for one phase of the gait cycle during normal walking.

Carter and associates[17] recently reported on data from several locations on the canine ulna and radius, and Lanyon and Baggot[16] reported on data from the cranial and caudal aspects of the radius of eight sheep. The results from studies such as these demonstrate that the in vivo stresses experienced by long bones during routine activities are oriented predominantly along the shaft of the bone. Stresses occurring perpendicular to the shaft are minimal. Based on this fact alone, the anisotropy exhibited by cortical bone appears to make sense teleologically.

Table 3
STRAINS AND STRESSES AT PUSH-OFF PHASE OF GAIT
CYCLE DURING NORMAL WALKING

			Ref.
Principal strains and orientation[a]	ϵ_I	= 395 μ strain	15
	ϵ_{II}	= 434 μ strain	
	θ	= −53°	
Strains in principal material directions	ϵ_3	= −134 μ strain	18
	ϵ_1	= 95 μ strain	
	γ_{13}	= −797 μ strain	
Principal stresses and orientation[a]	σ_I	= 2.06 MPa	18
	σ_{II}	= −3.67 MPa	
	θ	= −58°	
Stresses in principal material directions	σ_3	= −2.06 MPa	18
	σ_1	= 0.45 MPa	
	τ_{13}	= −2.57 MPa	

[a]   Orientation of principal strains and stresses given with respect to a proximally directed axis along the tibial shaft with postero-medial angles being positive.

IV. MULTIDIRECTIONAL FAILURE THEORIES

To explain the elastic behavior of bone and to interpret in vivo strain data required an anisotropic model, that of transverse isotropy. To predict the failure of bone, caused either by yielding or fracture, also requires a theory applicable to anisotropic, composite materials. The early studies of strength of anisotropic materials were performed primarily on wood and single crystals. In the wood industry, empirical interaction equations were developed to describe the uniaxial strength of wood as a function of grain orientation. Typical of these formulations is that due to Hankinson.[19] Early work on the anisotropic strength characteristics of crystals and textured metals was based on von Mises' isotropic yield criterion. Hill[20] generalized the von Mises criterion to account for material anisotropy. Azzi and Tsai[21] adopted the Hill yield criterion to develop a strength theory for composites capable of predicting strength under combined stresses. While this Tsai-Hill theory adequately describes off-axis strength data for composites and allows for normal stress interactions, it does not allow for differences in tensile and compressive strength known to exist for compact bone. The interaction effects are also described only in terms of the uniaxial strength in one of the principal material directions. To avoid these shortcomings, several variations of the Tsai-Hill theory have been proposed. Rosen,[22] for example, developed a failure criterion which describes interaction effects based on uniaxial strengths in two principal directions.

One major problem with the Tsai-Hill failure criterion (and its variations) is that they only apply for specifically orthotropic materials.[9] In order to allow for more accurate description of composite strength data and to provide a rational basis for strength transformations, Tsai and Wu[23] postulated a strength criterion in tensor polynomial form:

$$F_i\sigma_i + F_{ij}\sigma_i\sigma_j = 1 \tag{3}$$

where i, j = 1,2, . . . 6 and contracted tensor notation is used as in much the same manner as in Table 1 for the stress-strain relation. F_i and F_{ij} are second and fourth rank tensors, respectively. The linear terms in σ_i incorporate differences in tensile and compressive strengths. The quadratic terms in $\sigma_i\sigma_j$ allow all possible interactions between the stress components. The Tsai-Wu tensor has several advantages over the Tsai-Hill theory:

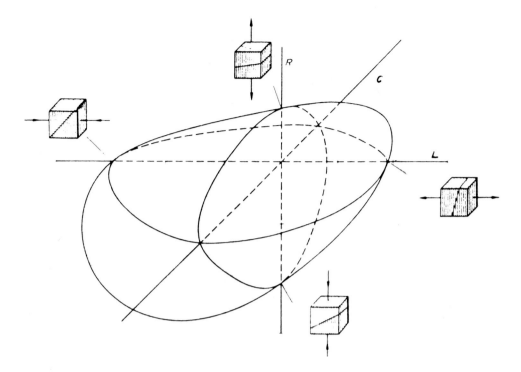

FIGURE 5.   Fracture locus for compact bone material. L, R, and C refer to stresses in the longitudinal, radial and circumferential directions, respectively. (From Pope, M. D. and Outwater, J. O., *J. Biomech.*, 7, 65, 1974. With permission.)

1.   It is invariant under rotation of the coordinate system.
2.   It can be transformed using existing tensor transformation laws.
3.   It has the same symmetry properties as the stiffness and compliance matrices for the material.

   Tsai-Hill theory and its variations have been previously applied in several off-axis strength studies of compact bone. Margel-Robertson[24] employed the Tsai-Hill criterion to describe the compressive behavior of laminated bovine bone. Compression tests were performed on standardized specimens oriented at various angles with respect to the long axis of the bone. The experimentally measured compressive strengths vs. angle of orientation were compared to theoretical predictions and good agreement was observed. Pope and Outwater[25] applied the Rosen variation of the Tsai-Hill theory to bovine Haversian bone. Bending tests were conducted on standardized specimens to determine the mechanical properties as a function of orientation. The variation in measured strength with orientation was in good agreement with that predicted by the Tsai-Hill theory. With their results and results in the literature they postulated a fracture locus for bone, shown in Figure 5. More recently, Tateishi and co-workers[26] reported off-axis tensile tests on bovine compact bone. They employed the Tsai-Hill theory to describe the test data by choosing uniaxial strengths which gave a best fit between predicted and experimental values. Again, Tsai-Hill theory gave an adequate description of the experimentally determined variation in strength with orientation.

   Reilly and Burstein,[11] in their study of elastic and ultimate properties of compact bone, collected considerable off-axis strength data for both bovine and human Haversian bone in both tension and compression. For both the longitudinal and transverse (i.e., radial and circumferential) directions, they found the ultimate compression strength to be significantly higher than the ultimate tensile strength (Table 4). The longitudinal strengths in both tension

Table 4
ULTIMATE STRENGTH OF HUMAN
FEMORAL CORTICAL BONE IN THE
DIRECTIONS OF MATERIAL SYMMETRY

Longitudinal	Tension	σ_{ult} = 133 MN/m^2
	Compression	σ_{ult} = 193 MN/m^2
Transverse	Tension	σ_{ult} = 51 MN/m^2
	Compression	σ_{ult} = 133 MN/m^2

Data from Reilly, D. T. and Burstein, A. H., *J. Biomech.*, 8, 393, 1975. With permission.

and compression were significantly higher than the transverse strengths. The Hankinson criterion was used to describe the off-axis ultimate strengths, and predictions from this criterion were compared to the experimental values from off-axis tests. A typical example is shown in Figure 6. They concluded that this criterion effectively described the off-axis experimental results in both compression and tension for specimens taken from the plane of the longitudinal and circumferential directions.

More recently, Hayes and Wright[27] and Cowin[28] have independently proposed a strength theory for compact bone based on that of Tsai and Wu as described in Equation 3. As with other anisotropic failure criteria, the Tsai-Wu theory is empirical and is not meant to explain failure mechanisms or to predict fracture planes. Rather, it is meant to allow structural failure predictions based on known strength quantities. For the general anisotropic case, the strength tensors F_i and F_{ij} in the failure criterion contain 6 and 21 components, respectively, and are similar in form to tensors describing the elastic properties of anisotropic materials. From the symmetry properties of a transversely isotropic material such as bone, the tensors F_i and F_{ij} contain two and five independent components, respectively. By specializing the failure criterion to plane stress in the longitudinal-transverse plane, Equation 1 becomes:

$$F_1\sigma_1 + F_2\sigma_2 + F_{11}\sigma_1^2 + F_{22}\sigma_2^2 + F_{66}\sigma_6^2 + 2F_{12}\sigma_1\sigma_2 = 1 \qquad (4)$$

The components F_1, F_2, F_{11}, F_{22}, and F_{66} are easily determined from uniaxial tension and compression tests along the material axes of symmetry and from a pure shear test. Therefore, except for F_{12}, the components of the Tsai-Wu criterion for plane stress can be determined from existing bone strength data.

The determination of the interaction term, F_{12}, requires a combined state of stress. A combined state of stress exists in a composite material-like bone whenever the applied stress is not coincident with the material axes of symmetry. Though sufficient biaxial strength data to calculate F_{12} do not presently exist, values can be determined from approximations and from limits defined by the stability conditions inherent in the theory. Hayes and Wright[27] used these limits to show agreement between the Tsai-Wu approach and existing off-axis bone data as shown in Figure 7. The investigation of bone properties in off-axis or biaxial tests and the development of an anisotropic strength theory to incorporate the results represent the first step in relating multiaxial strength data to the structural characteristics of whole bones.

V. CONCLUDING REMARKS

Cortical bone is an anisotropic material displaying different mechanical properties in different directions. The composite nature of cortical bone can be seen on microscopic and macroscopic levels by examining the tissue morphologically. As it comprises the tubular

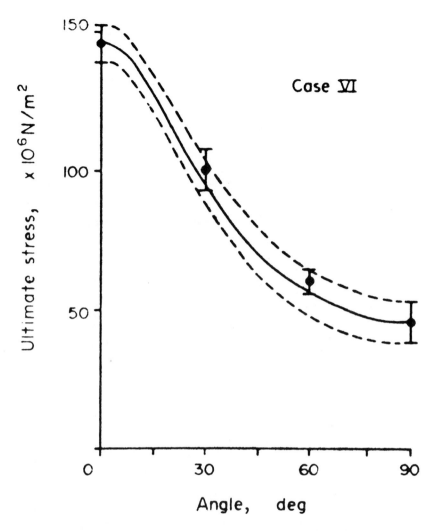

FIGURE 6.   Off-axis strength in tension for bovine Haversian femoral specimens. (From Reilly, D. T. and Burstein, A. H., *J. Biomech.*, 8, 394 and 402, 1975. With permission.)

shafts of the load-bearing long bones of the body, cortical bone can be modeled as a transversely isotropic material. The predominant material direction in terms of strength and modulus of elasticity is aligned with the long axis of the bone. When supported by in vivo measurements which show maximum stresses being applied in this direction, the anisotropy would appear justified teleologically. Even so, stresses resulting in cortical bone tissue are often complex. To gain understanding of bone overload and fracture requires an anisotropic failure theory supported by experimental strength data from off-axis and biaxial tests.

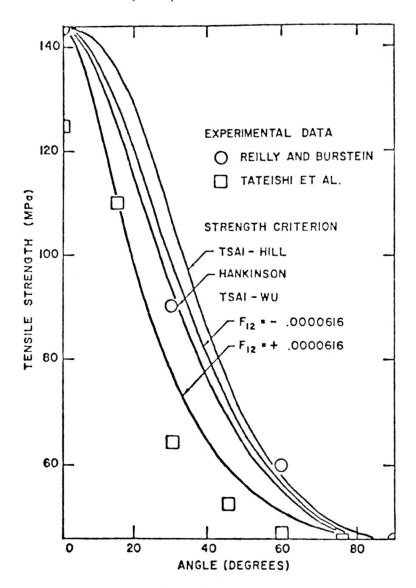

FIGURE 7.   Off-axis tensile strength for compact bone. (From Hayes, W. C. and Wright, T. M., *Advances in Research on the Strength and Fracture of Materials,* Vol. 3B, Taplin, D. M. R., Ed., Pergamon Press, New York, 1977, 1178. With permission.)

REFERENCES

1. **Carter, D. R., Hayes, W. C., and Schurman, D. J.,** Fatigue life of compact bone. II. Effects of microstructure and density, *J. Biomech.,* 9, 211, 1976.
2. **Smith, J. W.,** Collagen fiber patterns in mammalian bone, *J. Anat.,* 94, 329, 1960.
3. **Enlow, D. H.,** An evaluation of the use of bone histology in forensic medicine and anthropology, in *Studies on the Anatomy and Function of Bones and Joints,* Evands, F. G., Ed., Springer-Verlag, New York, 1966, 93.
4. **Enlow, D. H.,** The functional significance of the secondary osteon, *Anat. Rec.,* 142, 230, 1962.

5. **Pritchard, J. J.,** General morphology of bone, in *The Biochemistry and Physiology of Bone,* Vol. 1, Bourne, G. H., Ed., Academic Press, New York, 1972, 1.
6. **Dempster, W. T. and Liddicoat, R. T.,** Compact bone as a non-isotropic material, *Am. J. Anat.,* 91, 331, 1952.
7. **Reilly, D. T. and Burstein, A. H.,** The mechanical properties of cortical bone, *J. Bone Jt. Surg.,* 56a, 1001, 1974.
8. **Lang, S. B.,** Ultrasonic method for measuring elastic coefficients of bone and results on fresh and dried bovine bone, *IEEE Trans. Biomed. Eng.,* BME-17, 101, 1970.
9. **Jones, R. M.,** Mechanical behavior of lamina, in *Mechanics of Composite Materials,* McGraw-Hill, New York, 1975, 37.
10. **Yoon, H. S. and Katz, J. L.,** Ultrasonic wave propagation in human cortical bone. II. Measurements of elastic properties and microhardness, *J. Biomech.,* 9, 459, 1976.
11. **Reilly, D. T. and Burstein, A. H.,** The elastic and ultimate properties of compact bone tissue, *J. Biomech.,* 8, 393, 1975.
12. **Cochran, G. V. B.,** Implantation of strain gauges on bone in-vivo, *J. Biomech.,* 5, 119, 1972.
13. **Roberts, V. L.,** Strain-gauge techniques in biomechanics, *Exp. Mech.,* 6, 1, 1966.
14. **Lanyon, L. E.,** Analysis of surface bone strain in the calcaneus of sheep during normal locomotion, *J. Biomech.,* 6, 41, 1973.
15. **Lanyon, L. E., Hampson, W. G. J., Goodship, A. G., and Shan, J. S.,** Bone deformation recorded in vivo from strain gages attached to the human tibial shaft, *Acta Orthop. Scand.,* 46, 256, 1975.
16. **Lanyon, L. E. and Baggott, D. G.,** Mechanical function as an influence on the structure and form of bone, *J. Bone Jt. Surg.,* 58b, 436, 1976.
17. **Carter, D. R., Smith, D. J., Spengler, D. M., Daly, C. H., and Frankel, V. H.,** Measurements and analysis of in-vivo bone strains on the canine radius and ulna, *J. Biomech.,* 13, 27, 1980.
18. **Carter, D. R.,** Technical note: anisotropic analysis of strain rosette information from cortical bone, *J. Biomech.,* 11, 199, 1978.
19. **Hankinson,** Investigation of Crushing Strength of Spruce at Varying Angles of Grain, U.S. Air Service Information Circular No. 259, 1921.
20. **Hill, R.,** Plastic anistrophy, in *The Mathematical Theory of Plasticity,* Clarendon Press, Oxford, 1950, 317.
21. **Azzi, V. D. and Tsai, S. W.,** Anisotropic strength of composites, *Exp. Mech.,* 5, 283, 1965.
22. **Rosen, B. W.,** Mechanics of Composite Strengthening, Space Science Lab Report No. R64SD80, General Electric, 1964.
23. **Tsai, S. W. and Wu, E. M.,** A general theory of strength for anisotropic materials, *J. Compos. Material,* 5, 58, 1971.
24. **Margel-Robertson, D. R.,** Studies of Fracture in Bone, Ph.D. dissertation, Stanford University, Stanford, California, 1973, 155.
25. **Pope, M. D. and Outwater, J. O.,** Mechanical properties of bone as a function of position and orientation, *J. Biomech.,* 7, 61, 1974.
26. **Tateishi, T., Nonaka, K., Shiraski, Y., Kimura, T., and Ogawa, K.,** Mechanical properties of bovine femur and its fractography, *Bull. Mech. Eng. Lab.,* Japan, No. 21, 1976.
27. **Hayes, W. C. and Wright, T. M.,** An empirical strength theory for compact bone, in *Advances in Research on the Strength and Fracture of Materials,* Vol. 3B, Taplin, D. M. R., Ed., Pergamon Press, New York, 1977, 1173.
28. **Cowin, S. C.,** On the strength anisotropy of bone and wood, Paper 79-WA/APM-21, paper presented at Winter Annual Meeting, American Society of Mechanical Engineers, New York, 1979.

Chapter 4

PRINCIPLES AND METHODS OF SOLID BIOMECHANICS

R. Huiskes

TABLE OF CONTENTS

I.　　Introduction...52

II.　　Rigid Body Biomechanics 53

III.　　Stresses and Strains..59
　　　A.　　Introduction and Definitions..59
　　　B.　　Mechanical Properties of Materials.....................................63

IV.　　Structural Stress Analysis...67
　　　A.　　Introduction 67
　　　B.　　Methods of Analysis 68
　　　　　1.　　Experimental Methods 68
　　　　　2.　　Closed-Form Theories 72
　　　　　　　a.　　Bar Theory..75
　　　　　　　b.　　Compound-Bar Theory 76
　　　　　　　c.　　Beam Theory...78
　　　　　　　d.　　Torsion of Circular Shafts...............................82
　　　　　　　e.　　Combined Loading of Slender Bodies.....................85
　　　　　3.　　Finite Element Methods..88

Appendix: Parameters and Units 95

References..96

I. INTRODUCTION

Biomechanics entails the application of methods and principles of engineering mechanics to biological structures and medical problems. As such, it is not a new field of endeavor. Early efforts in this area date back to Aristotle, Leonardo da Vinci, and Galileo. Significant contributions to the understanding of human body mechanics were also made in the last century and the first half of this century. Biomechanics has many fields of application, including orthopedic and cardiovascular surgery, traumatology, dentistry, rehabilitation, and sports, and is closely interwoven with basic medical sciences such as biophysics and medical physics, physiology, functional anatomy, and biomaterials. It can be considered a subbranch of biomedical engineering (or bioengineering) and a branch of biomechanical engineering, as engineering mechanics is one basic science of mechanical engineering.

After the World War II and during the last two decades specifically, a proliferation in biomechanics activities has occurred. This development was triggered and enhanced by two separate mechanisms. In the first place, engineering mechanics has advanced tremendously through the development of computers and computer methods. As a result, the complex mechanical behavior of biological tissues and structures can now be realistically described and successfully analyzed. Secondly, increasing emphasis is being put on close surgical reconstruction of body functions in the disabled and the sick. The augmentation in routine applications of artificial joints and new fracture fixation devices in orthopedic surgery is a good example. Long-term success of these devices and other reconstructive operations requires designs and surgical techniques that are based on a sound understanding of human body musculoskeletal mechanics.

Biomaterials are usually applied in devices that carry load and transfer it to their biological environment, in orthopedic surgery and dentistry specifically. It is evident that the shape of these devices and the properties of their materials should be such that they are able to carry the loads without failure for an adequate period of time. However, in addition, adverse biological reactions of tissues to implant devices, due to mechanical causes, have to be taken into account as well because the integrity and biological behavior of tissues depend on their mechanical, as well as on their chemical and electrical environment. In other words, laboratory testing of biomaterials with respect to strength and deformational behavior is really not enough when these materials are to be applied in load-carrying (and load-transmitting) implant devices. Thorough mechanical analyses of the implant-tissue composites are equally necessary in a rational approach to prosthetic design. The mechanics methods and principles needed for such an approach are difficult because of the complex nature of the (biological) structures concerned. This chapter can only touch upon the basics, that may nevertheless be adequate for the layman to appreciate the aims and scopes of biomechanics research, and judge the validity and significance of results at least to some extent.

Engineering mechanics can be divided into rigid body mechanics (dynamics, kinematics, and statics), fluid mechanics, and solid mechanics. In view of the scope of this book and the space available, we will limit ourselves principally to solid biomechanics of hard tissues. After a brief discussion of force analysis in rigid body mechanics, the concepts of stress and strain are reviewed and the way in which materials react to load discussed, focusing on definitions of terminology. In the last section, the philosophy, principles, and methods of stress analysis are treated. This subject was thought to be the most appropriate for the scope of this volume; the preceding sections merely introduce some basic principles.

Mechanics cannot be explained nor used without the introduction of parameters that express properties and phenomena in quantifiable form. A list of parameters and their units as used in the text is provided in the Appendix. The readership of this chapter was anticipated as consisting of (relatively) uninformed, but interested scientists. It will be evident that for such a group a full and in-depth treatment of solid biomechanics would not fit into the concept of the book. Nor is it realistic to expect that this chapter could provide all the knowledge needed to understand the remainder of the volume. Whenever a compromise had to be made, the choice was for generality: it was tried to set up a framework in which stress analysis of structures takes place, providing and explaining the essential terminology, and discussing the philosophy, occasionally touching upon some examples. The holes in the framework must, by necessity, be filled elsewhere. In the References, a list of suggested reading material is presented.

II. RIGID BODY BIOMECHANICS

Forces in the musculoskeletal system are generated by gravity, muscle action, acceleration and deceleration of body parts, and joint restraints. These forces cause motion in the system and deformation of skeletal parts. Because the deformations are small when compared to the gross motions, they are usually neglected when forces and gross motions are analyzed. In other words, the skeletal parts are assumed rigid in that case.

The study of forces and motions between rigid bodies is called kinetics, and the analysis of motions alone is called kinematics. If a rigid body is at rest, or moving with a constant speed (no accelerations), its external forces form an equilibrium system. The study of this condition is called statics. The principles of statics are formulated in the laws of Newton.

It is not the aim of this section to provide for an in-depth treatment of statics, but just to briefly discuss a few aspects that are of importance in the identification and evaluation of forces on human body parts. Consider the example shown in Figure 1. Let us assume that we wish to evaluate the ankle joint and Achilles tendon forces in the foot during gait. In order to facilitate our analysis, a number of simplifying assumptions are made:

1. Although the structure to be analyzed (the foot) is not at rest and probably not moving at a constant speed, we assume that acceleration forces can be neglected when compared to muscle, joint restraint, and floor reaction forces (in other words, the problem is assumed quasistatic). For the same reason we neglect gravity forces.
2. We regard the foot in subsequent positions of gait; in each position (and we will consider only one here) we assume the foot to be a rigid body, taking into account its specific configuration (geometry) at that time.
3. We assume all forces to work in one plane. Owing to these three assumptions we can now apply two-dimensional (2-D) statics.

The next step is to identify all relevant external forces working on the body. For that reason the body (the foot in the position shown in Figure 1) is freed from its environment, replacing all environmental restraints by forces. Forces are vector quantities and are characterized by orientation, magnitude, and point of application. As a reference we define an x-y coordinate system, indicated in Figure 1. We expect external forces on the foot at three locations: the Achilles tendon insertion, the ankle joint, and the foot-floor contact region. Of the tendon force (F_t) we know the orientation as well as the point of application (from X-rays), not the magnitude. Of the ankle joint force (F_a), we know the point of application only.* Of the ground reaction force (F), we know the point of application, not the orientation, nor the magnitude. Thus, we have five unknowns: the magnitudes of F, F_t, and F_a and the orientations of F and F_a. Since each force can be decomposed in vector components, we can change the nature of the unknowns. For example, the ground reaction force can be decomposed in a force in the x direction (F_x) and a force in the y direction (F_y), a force

---

* Because articular joints are almost frictionless, the line of application of the joint forces must by necessity cross the (instantaneous) joint rotation center (see also Figure 2).

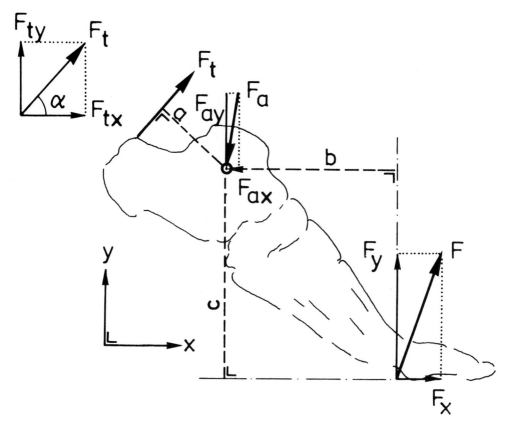

FIGURE 1.   A free body diagram of the foot in a specific position. Achilles tendon, ankle, and floor reaction forces are drawn and decomposed in their x and y components. Moment arms (a, b, c) to the joint instantaneous center of rotation are shown.

system that is equivalent to the original force F (which is the vector sum of F_x and F_y). In the same way F_a is decomposed in F_{ax} and F_{ay}. We still have five unknowns (the magnitudes of F_t, F_x, F_y, F_{ax}, and F_{ay}), but of a different nature. The entity of the body under consideration, freed from its environment, in which the environmental restraints are represented by forces (as shown in Figure 1) is called a free body diagram.

The last step is to evaluate the unknown forces by applying the equilibrium conditions: a body is in equilibrium when the vector sum of all external forces is zero and when the sum of the moments* of all forces with respect to one arbitrarily chosen point is zero.

This condition gives us three equations in a 2-D case:

1.   Sum of forces in x direction = 0
2.   Sum of forces in y direction = 0
3.   Sum of force moments = 0

In order to apply the first two conditions to our example, F_t must be decomposed in the x and y directions as well. From the geometric configuration it follows: $F_{tx} = F_t \cos\alpha$ and $F_{ty} = F_t \sin\alpha$, in which α is known. Applying the equilibrium conditions we find:

$$1.\ F_x - F_{ax} + F_t\cos\alpha = 0$$

* Moment of a force with respect to a point = force magnitude × moment arm (the shortest distance between the point and the line of action of the force), clockwise positive and anticlockwise negative.

2. $F_y - F_{ay} + F_t \sin\alpha = 0$
3. $F_t a - F_y b - F_x c = 0$

Since we have five unknowns and only three equations, we cannot solve this system unless two unknowns are eliminated by other means. Such a force system is called indeterminate. A method of obtaining more information is to measure magnitude and orientation of the ground reaction force F (and thus F_x and F_y), for instance, by using a force plate. We then reduce the number of unknowns to three, which can be evaluated from the three equations.

It is evident that this example is a simple one, at least we made it a simple one by introducing a number of assumptions. In reality the problem may be three-dimensional (3-D), while a number of additional muscle tendons and ligamentous restraints may load the body as well. Dynamic aspects may play a role too. In other words, the solution found in this way cannot be more than a rough estimate.

Nevertheless, the example serves to illustrate a number of aspects that are of importance in biomechanics force analysis:

● The introduction of specific assumptions about the mechanical behavior of the system considered, in order to facilitate the analysis or even make it possible (These assumptions must, however, be realistic with respect to the required validity of the results.)

● The development of the free body diagram by identifying the body under consideration and isolating it from its environment, replacing the environmental restraints by forces

● The application of geometric data (for instance from X-rays) and the nature of connections to determine as many characteristics of the external forces as possible (We recognized the nature of the tendon insertion to estimate the point of application and the orientation of the Achilles tendon force. We used the nature of the ankle joint to determine the point of application of the ankle force. A characterization of several types of connections and their specific force systems is shown in Figure 2.)

● The application of equilibrium conditions to calculate the unknowns; three equations in a 2-D problem, six in a 3-D problem* (Vector decomposition of forces in the directions of the coordinate axes is usually necessary.)

● The fact that the force system may be indeterminate, so that unknowns will have to be eliminated by other means, for instance by using experimental data or more sophisticated mathematical criteria

The last aspect is a problem in musculoskeletal biomechanics research specifically.[1] The number of muscles and joint/ligamentous restraints working on the individual bones is large, while the experimental means available to measure these forces are of limited applicability and accuracy. Electromyography can be used to estimate muscle action, but only to a limited extent. Force plates can be applied to evaluate ground reaction forces on the foot, and center-of-gravity analysis can be used for gravity force determination on body parts. In dynamic performances, for instance, in gait, acceleration forces on body parts can be calculated by rough approximation from motion patterns. However, the experimental methods and the equilibrium equations available are not sufficient in most cases to obtain a complete and dependable estimate of all muscle and joint forces. Additional mathematical tools, as for instance, mathematical optimization criteria, can be applied. When it is assumed that the forces generated in muscles and joints are such that the total muscle energy expenditure is

---

* In three dimensions: three force equations for the x, y, and z directions and three moment equations for the x-y, y-z, and the z-x planes.

connection   symbol   unknowns

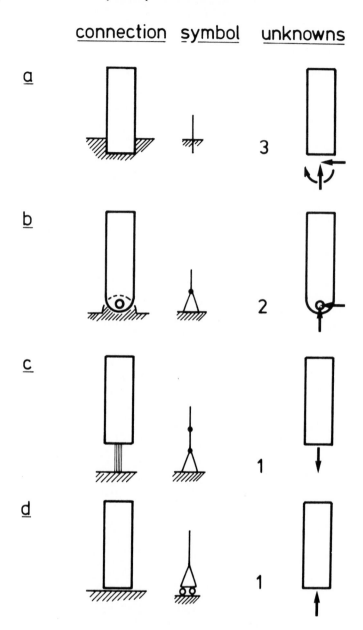

FIGURE 2.   Characterization of unknown loading variables in kinematic connections of structures to their environment. (a) Rigid fixation (two forces and a moment); (b) hinge (two forces or one force with unknown orientation); (c) tendon or cable (one force in the direction of the tendon); and (d) frictionless sliding (one force perpendicular to the sliding surfaces).

minimal during a specific function, additional mathematical equations are obtained. The question of course is whether this criterion is realistic. Alternative criteria have been used as well: minimal joint forces, optimal control function, and others; the results usually differ for each criterion chosen. Altogether, the present knowledge about forces in human joints is of a rough approximate nature and limited to a few specific functions, as for instance, gait. Evidently, a rational approach to artificial joint design by the application of mechanics analysis is hampered by this lack of quantitative knowledge.

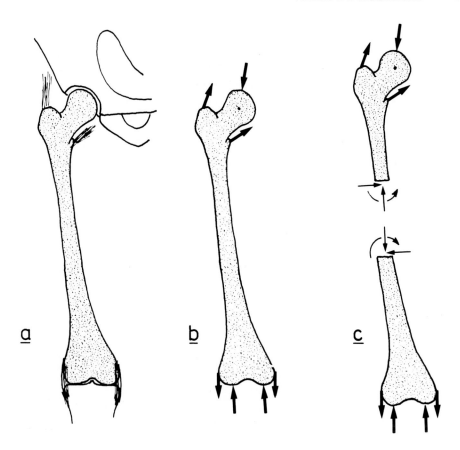

FIGURE 3. (a) A femur developed in (b) a free body diagram, and (c) free body diagrams of two parts of that bone. The principle of cutting is applied in the latter case by introducing a loading system of two forces and a moment at the site of the (imaginary) cut.

Developing the free body diagram is a very important step in all problems of force and stress analysis. All subsequent steps depend on this diagram; errors made here will be reflected in the validity of the eventual conclusions. The crucial aspect of this step is freeing the body from its environment and replacing the connections by forces and moments to take their influences into account. We have seen (in Figure 2) how kinematic connections, restraints, can be represented by forces and couples. The body under consideration, however, the structure to be analyzed, is not by necessity a physical entity. To analyze a part of a body, we apply the principle of cutting, illustrated in Figure 3. A free body diagram of the femur (see Figure 3a) is developed by freeing this bone from its environment and replacing all environmental connections by forces, taking into account the kinematic characteristics of the restraints (see Figure 3b). A free body may comprise only a part of the bone as well, for instance, if the bone is cut as shown in Figure 3c. The mechanical influence of the removed part must be taken into account, just as in the case of freeing the body from environmental connections, by introducing forces and couples. These forces and couples represent the internal loads within the material, at the site of the cut. Due to the action = reaction law, the forces and couples must be equal but opposite in sign on either side of the cut. The nature of the sectional loading system is equal to that of a rigid fixation (two forces and a couple in a 2-D problem, see Figure 2).

The sectional, internal loads are used to free a part of a body in a free body diagram for a subsequent force analysis, in which these loads are evaluated. Let us for instance consider

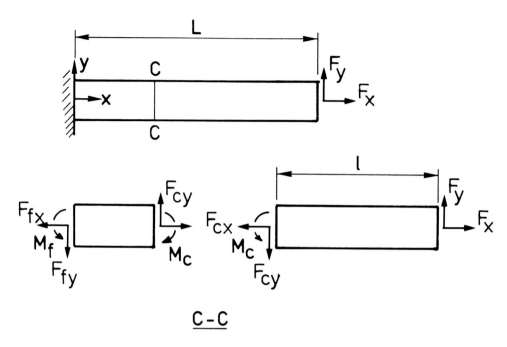

FIGURE 4.   Internal loads in a cantilever beam, loaded externally by a transverse and an axial force. The internal forces and moments at an arbitrarily chosen site can be calculated using equilibrium conditions.

the cantilever beam in Figure 4, loaded by an axial (F_x) and a transverse (F_y) force. The beam is first freed from its fixation by introducing the forces F_{fx} and F_{fy} and the moment M_f. Applying equilibrium conditions we find:

$$\text{x direction: } F_x - F_{fx} = 0, \text{ hence } F_{fx} = F_x$$

$$\text{y direction: } F_y - F_{fy} = 0, \text{ hence } F_{fy} = F_y$$

$$\text{moments: } M_f + F_y L = 0, \text{ hence } M_f = -F_y L$$

Apparently, M_f works in the opposite direction as drawn, as witnessed by the negative sign. If we now wish to evaluate the internal loads in the beam at cross section CC, the beam must be cut at that location and the cross-sectional loads F_{cy}, F_{cx}, and M_c introduced. The reader can verify by applying equilibrium conditions, that $F_{cx} = F_x$, $F_{cy} = F_y$, and $M_c = -F_y \ell$. If this is carried out for varying ℓ, the internal load distribution is obtained, giving the cross-sectional loads throughout the beam in diagram, as shown in Figure 5. Evidently, the internal axial and transverse forces are constant throughout the beam, while the internal moment varies from zero to maximal at the fixation site. These force and moment diagrams are often applied for beam-like structures for design and analysis purposes. It is evident in this example that the greatest structural strength is required at the fixation site.

Another example of this procedure is worked out in Figure 6, with respect to the femoral shaft. In this problem it is assumed that the abductor muscles are active only and that all forces work in one plane (2-D problem). The analysis is carried out in steps, starting from the free body diagram of the femur (in which all knee forces are lumped into three loading variables):

1.   Decompose the external forces in x and y directions.

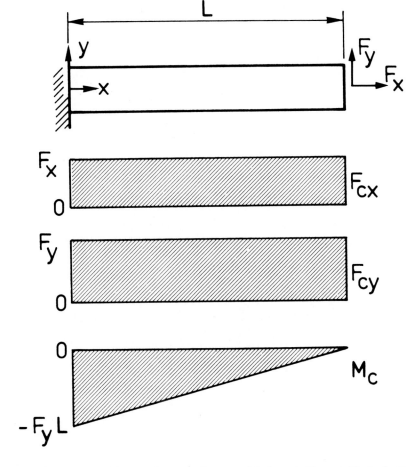

FIGURE 5.   Internal forces and moment diagrams of the beam in Figure 4. The loads are shown as functions of the longitudinal coordinate x. The internal forces are constant throughout the beam; the internal moment varies linearly, with a maximum in absolute value at the fixation site.

2.   Develop a free body diagram of the femoral shaft by evaluating the loads at Sections A-A and B-B.
3.   Calculate the cross-sectional loads in the shaft as a function of ℓ.
4.   Draw the diagram for the internal load distribution, based on Step 3.

III. STRESSES AND STRAINS

A. Introduction and Definitions

When a body, a piece of material, is subject to loading, it deforms. Although the deformation may be invisible to the naked eye, it is always present. The molecules resist deformation through mutual bonds that generate internal loads. In solid mechanics, which studies this deformational behavior under loading, the local deformations are represented by strain; the internal loads are represented by stress. In the analyses of these phenomena in structures (called structural analysis), the principles of continuum theory are usually applied. Although the materials out of which a structure is made are not truly continuous at a molecular or even a microscopic level, the mechanical behavior of structures can usually be described

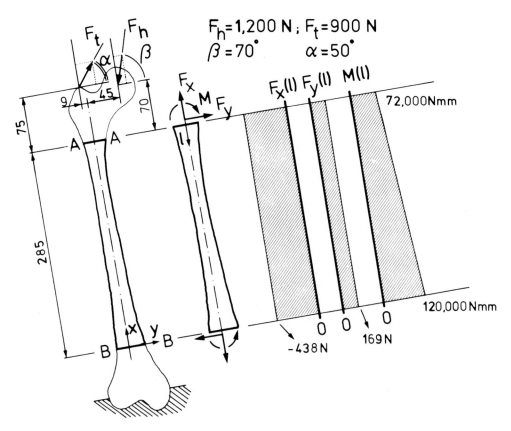

FIGURE 6.   Internal forces and moment diagrams of a femoral diaphysis. All forces are assumed to act in one plane. Dimensions are shown in millimeters. From the abduction tendon force magnitude (F_t) and direction (α), the hip force magnitude (F_h) and direction (β), and the geometric configuration, the internal loads at section AA (F_x, F_y, and M) are calculated using equilibrium conditions. These are the external loads of the freed bone piece AA - BB in which, just as in Figure 5, internal forces and moment as functions of the location (ℓ) are calculated.

and predicted with methods that neglect discontinuities on a smaller scale. Within the frame of continuum theory, materials are regarded as indefinitely divisible and thus variables as stresses and strains can be defined in an indefinitely small point. This assumption of material continuity is sometimes quite adequate, as in metals, and sometimes rather rough, as for instance, in trabecular bone.

Strain is relative deformation (a change in dimension relative to the original dimension); stress is an amount of force per unit area; both are local phenomena. To define the stress state in a point of a structure, we chose an arbitrary plane in that point (see Figure 7a). On this plane works a certain internal load that, divided by the plane area, is represented by a certain amount of stress (σ). The stress can be decomposed into a component normal to the plane (σ_n, called direct stress) and one parallel to the plane (τ, called shear stress). Thus, the stress state in that point is fully characterized by three variables: one direct-stress component (σ_n), one shear-stress component (τ), and the orientation (α) of the plane with respect to an external (x-y) reference system. This characterization of stress in a point can be worked into two alternative representations that are commonly used. In the first one (see Figure 7b), two planes are chosen parallel to the external coordinate axes (an x-plane and a y-plane). On these plates we find again direct-stress components (σ_x and σ_y) and shear-stress components (τ_{xy} and τ_{yx}).* It can be proven that this system of stress components on nonarbitrarily

---

*   The convention used is as follows: σ_x is a direct stress on a plane perpendicular to the x axis (an x plane), pointing in the positive x direction (σ_y and σ_z, accordingly). τ_{xy} is a shear stress on an x plane in positive y direction, etc.

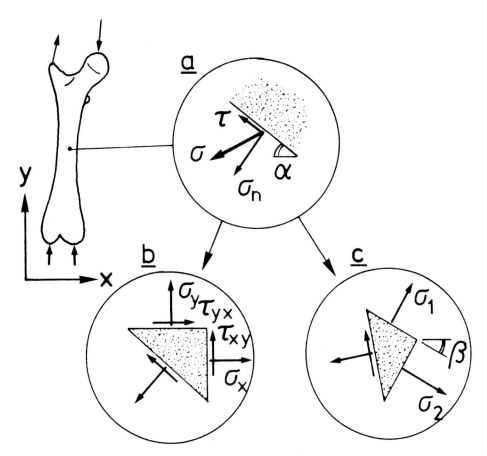

FIGURE 7. Characterization of a 2-D stress state. (a) In a geometric point by three variables (direct stress, shear stress components, and plane orientation); (b) alternative characterization with respect to fixed planes; and (c) by principal stresses.

chosen planes is equivalent with that of Figure 7a and also that $\tau_{xy} = \tau_{yx}$. The third representation (see Figure 7c) is based on the fact that two perpendicular planes can be found on which only direct stresses exist (σ_1 and σ_2). These are called the principal stresses in the point concerned, while the orientation of the planes is characterized by the principal-stress orientation (β) with respect to the external coordinate system. In summary, the stress state in a point of a structure is characterized by either the magnitudes of two direct stress and one shear-stress component (σ_x, σ_y, τ_{xy}) or by the magnitudes of two principal-stress components and the principal-stress orientation (σ_1, σ_2, and β). Note that we did not discuss compressive and tensile stress. Both are direct stresses, compressive denoted with a negative sign and tensile with a positive sign. Although their influences on materials might differ, mathematically they are not treated separately.

Like a 2-D stress state (the example above) is characterized by three stress components, a 3-D stress state is characterized by six: the magnitudes of three direct-stress and three shear-stress components. Of course, all structures are in fact 3-D, but quite often the stress state can be adequately represented by a 2-D approximation. The six components of a 3-D stress state are illustrated in Figure 8. Parallel to the representation of a 2-D stress state as in Figure 7b, we regard an infinitesimal small cube in the material, the sides of which align with the external (x-y-z) reference system. On each of the three planes work one direct-stress and two shear-stress components. In this case, too, it can be proven that $\tau_{xy} = \tau_{yx}$, $\tau_{xz} = \tau_{zx}$, and $\tau_{yz} = \tau_{zy}$. Hence, the stress state in the point concerned is fully characterized by six independent variables, three direct- and three shear-stress components.

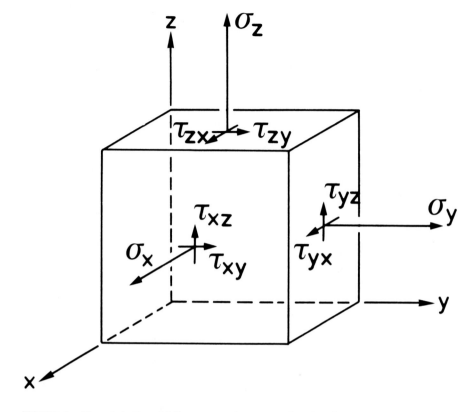

FIGURE 8.   Characterization of 3-D stress state in a geometric point with respect to fixed (x, y, z) planes in nine stress components. Because $\tau_{xy} = \tau_{yx}$, etc., the number of independent stress variables is reduced to six.

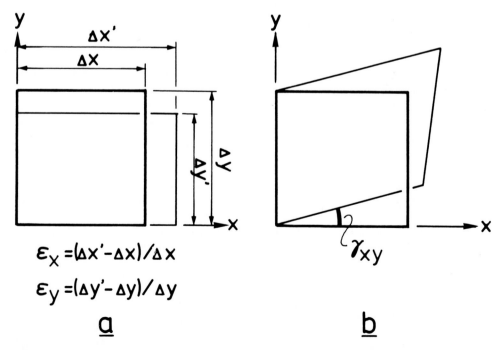

$$\varepsilon_x = (\Delta x' - \Delta x)/\Delta x$$

$$\varepsilon_y = (\Delta y' - \Delta y)/\Delta y$$

<u>a</u>

<u>b</u>

FIGURE 9.   Illustration of strain components in two dimensions. (a) The direct strains (ϵ_x, ϵ_y) represent relative dimensional changes of the infinitesimal cube; (b) the shear strain ($\gamma_{xy} = \gamma_{yx}$) represents the distortion of the cube.

The local state of deformation in a material is represented by strain, in much the same way as internal loading is represented by stress, and defined again with respect to an infinitesimal small cube, a geometric point. There are two types of strain (Figure 9): direct strains (ϵ), describing the relative changes in length of the cube ribs, and shear strains (γ), describing the angular distortion. Comparable to the stress state, the local strain state in a material is fully determined by six independent strain variables (i.e., ϵ_x, ϵ_y, ϵ_z, γ_{xy}, γ_{xz}, γ_{yz}) in a 3-D case and by three (i.e., ϵ_x, ϵ_y, γ_{xy}) in a 2-D case.

The stress and strain states in a point of a structure are related to each other through the mechanical (elastic) properties of the materials concerned. These relations are expressed in constitutive equations which mathematically relate the stress to strain components. The forms of these equations depend on the nature of the material, as discussed in the next section. Their simplest forms are for linear elastic, isotropic, and homogeneous materials * as metals,

$$\epsilon_x = \{\sigma_x - \nu(\sigma_y + \sigma_z)\}/E$$

$$\epsilon_y = \{\sigma_y - \nu(\sigma_x + \sigma_z)\}/E$$

$$\epsilon_z = \{\sigma_z - \nu(\sigma_x + \sigma_y)\}/E$$

$$\gamma_{xy} = 2\,\tau_{xy}\,(1 + \nu)/E$$

$$\gamma_{xz} = 2\,\tau_{xz}\,(1 + \nu)/E$$

$$\gamma_{yz} = 2\,\tau_{yz}\,(1 + \nu)/E$$

in which E and ν are the elastic constants of the material concerned, the Young's modulus (or modulus of elasticity) and Poisson's ratio, respectively. These relations are also known as Hooke's law. It is evident that when the material properties are known (E and ν), the strains can be directly calculated from the stresses and vice versa, using these formulas. The expression $E/2(1 + \nu)$ appearing (in reversed form) in the equations is usually denoted with the symbol G and called the shear modulus of the material concerned.

Although the stress and strain state in a point of a structure is 3-D in general, in special cases simpler forms will occur. Examples are plane stress state (e.g., $\sigma_z = \tau_{xz} = \tau_{yz} = 0$), plane strain state (e.g., $\epsilon_z = \gamma_{xz} = \gamma_{yz} = 0$), and uniaxial stress state (e.g., $\sigma_y = \sigma_z = \tau_{xz} = \tau_{yz} = 0$). The first two are both examples of 2-D (Figure 10); the last one is an example of a 1-D stress state. Uniaxial stress state occurs for instance in a simple elastic bar in tension (Figure 11). Because the bar is long when compared to its thickness and unrestrained at its sides, we may assume that stress occurs in the longitudinal direction only (σ_x), all other stress components being zero. In addition we may assume that the stress is distributed homogeneously over the cross section. Hence, it follows directly from the formulas given above that the constitutive equations for this case reduce to:

$$\epsilon_x = \sigma_x/E, \quad \epsilon_y = -\nu\sigma_x/E, \text{ and } \quad \epsilon_z = -\nu\sigma_x/E$$

The stress can be calculated directly from the external load, $\sigma_x = F/A$, and the longitudinal strain can be calculated directly from the elongation, $\epsilon_x = \Delta\ell/\ell$. If both the external load and the elongation are measured, the Young's modulus (E) can be calculated from the formulas given, hence, a suitable test for determining properties of materials.

B. Mechanical Properties of Materials

Assume a material which is subject to a tensile test as described in the previous section

---

* These terms are explained in the next section.

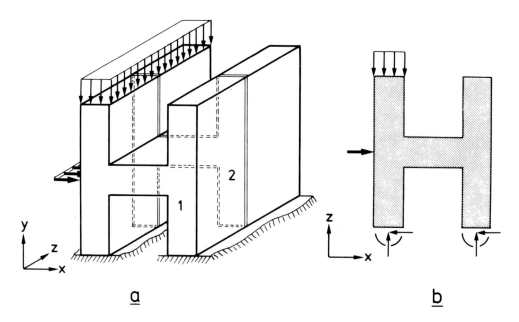

FIGURE 10.   (a) The loaded structure shown can be assumed plane because geometry and loading are equal in each section. (b) In the plane model either plane stress state ($\sigma_z = 0$) or plane strain state ($\epsilon_z = 0$) can be assumed. If the model represents Plane 1, then obviously $\sigma_z = 0$ and $\epsilon_z \neq 0$, because it is a free surface (free of stress, able to expand freely). If the model represents Plane 2, then $\epsilon_z = 0$ and $\sigma_z \neq 0$, because this plane is restrained in the z direction.

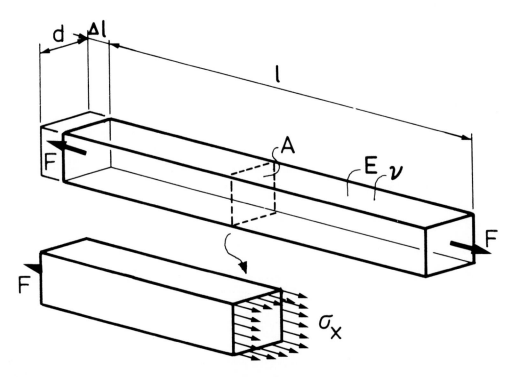

FIGURE 11.   A linear elastic, isotropic, and homogeneous bar with uniform cross section, loaded in tension. Uniaxial stress state presides and the stress is distributed uniformly over a cross section. The force engenders an elongation $\Delta\ell$.

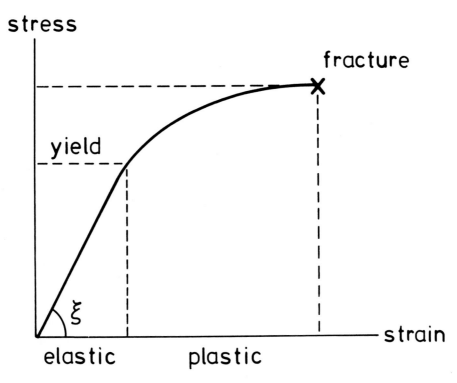

FIGURE 12. Stress-strain curve of an arbitrary material, tested as in Figure 11 by an increasing force. Fracture occurs when the stress reaches the value of the ultimate strength. In the first part of the curve, the deformation is elastic; in the second part it is plastic. The elastic stiffness is represented by the Young's modulus, the tangent of the angle ξ.

(see Figure 11). When the external load is increased from zero, the elongation of the bar increases until the material fractures. When the elongation and the force are continuously registered and calculated to strain (ϵ) and stress (σ), respectively, a so-called stress-strain curve is obtained (Figure 12). Because the stress and strain are variables relative to area and length, respectively, the stress-strain curve is independent of the test specimen dimensions, hence, characteristic of the material itself. This curve can be divided into an elastic region and a plastic region. A material stressed in the elastic region returns to its original configuration when the load is removed. The deformation obtained in the plastic region, however, is irreversible. Many materials (e.g., metals, bone) will display, at least in the lower stress-strain regions, linear elastic behavior, which means that strain is directly proportional to stress (contrary to nonlinear elastic behavior). The proportionality constant, the tangent of the angle between the straight part of the curve with the strain axis, is called the Young's modulus (E) of the material. The stress at which the material fractures is called the ultimate strength, the strain, the ultimate strain, or fracture toughness. The stress at which the material (approximately) enters the plastic region is called the yield strength. A material with a high Young's modulus is called stiff, contrary to flexible. A material with a high ultimate strength is called strong, contrary to weak. A material with a high toughness is called tough, contrary to brittle.

Knowledge about the elastic properties of materials, as expressed in their constitutive equations, is important for structural stress analyses, because these equations have to be applied in calculations. If a material displays equal deformational behavior when loaded in different directions, it is called isotropic, contrary to anisotropic. When a material is homogeneous, it has equal elastic properties at different locations, contrary to nonhomogeneous.

It should be kept in mind that all these material properties are defined for the material as such, on a local basis, irrespective of the geometry of the body or structure, within the concept of continuum theory. Some materials (e.g., collagenous tissues and, to a certain extent, plastics) display properties that are characteristic for viscous fluids. In such materials, called visco-elastic, the deformation depends not only on the loads, but on the loading rate (load history) as well. In other words, a stress-strain curve of such a material will alter for different loading rates. These materials are usually stiffer and stronger in fast loading and tougher in slow loading.

The simplest materials display linear elastic, isotropic, and homogeneous properties in their elastic region (e.g., metals). In that case, the constitutive equations can be fully characterized by a Young's modulus (E) and a Poisson's ratio (ν), as illustrated in the previous section.* We see from the constitutive equations for uniaxial stress state that $\epsilon_y = \epsilon_z = -\nu\epsilon_x$. Hence, Poisson's ratio can be interpreted as a compressibility constant, indicating how a material deforms in one direction if loaded in another. If $\nu = 0.5$, the total volume of the bar does not change after loading; such materials are called incompressible. The constitutive equations for nonlinear elastic, visco-elastic, and/or anisotropic materials are more complicated.

Plastics such as polyethylene and polymethylmethacrylate are visco-elastic to some extent, while acrylic cement such as that used for fixation of orthopedic implants is usually non-homogeneous in addition. Cortical and cancellous bone are known to be visco-elastic to some extent, anisotropic, and nonhomogeneous. Collagenous structures such as ligaments and joint cartilage are highly visco-elastic, due to their fluid contents, whereas they are anisotropic and nonhomogeneous, depending on the organization and distribution of their collagen fibers.

When elastic properties of materials are applied in stress analyses, their description with parameters in constitutive equations is always subject to simplifying assumptions. We have seen, for example, how we assumed materials to be continuous, although we know that they are not. The question is always whether a deviation from assumed behavior would significantly affect the conclusions of the analysis in the light of the analysis objectives. Let us, for instance, regard trabecular bone, which is macroscopically noncontinuous. Nevertheless, it is often assumed as being continuous in stress analyses. It is evident that stress predicted in such an analysis represents a rough lumped internal load only, when compared to stress as it would occur in the bone of the trabeculae themselves.[2] Another example is cortical bone, which is macroscopically continuous. Although cortical bone does display viscoelastic properties, its strain rate dependence is only slight and can be neglected in quasistatic analyses, if loading rates are not too fast.[3-5] Cortical bone is anisotropic and nonhomogeneous, but under certain conditions it can be assumed as transversely isotropic** and homogeneous and, depending on the required accuracy of an analysis, sometimes even fully isotropic.[3-5] These are a few examples that illustrate the principles of representing material properties in stress analyses by approximation.

The ultimate properties of materials are not applied in the stress analyses themselves, but to judge the significance of their results. The stresses predicted in a stress analysis are usually compared to a stress criterion to see what they actually indicate. Such a criterion is a quantitative representation of a failure mechanism. In the bar in tension discussed previously, for instance, failure mechanisms are irreversible (plastic) deformation and fracture. Their stress criteria are expressed in yield strength and ultimate strength. When the stress state is 2-D or 3-D, the (3 or 6) stress components must first be translated to an equivalent situation

---

* As shown previously, the ratio $G = E/2(1 + \nu)$ is called the shear modulus. So altogether, three parameters are commonly used to characterize linear elastic, isotropic, and homogeneous materials. Nevertheless, two of these are sufficient, as the third can always be derived.

** Transverse isotropy of cortical bone: the Young's modulus and Poisson's ratio in the longitudinal direction differ from those in the radial and circumferential directions, but are equal in the latter two.

of uniaxial stress state. In that case, the stress components are calculated to what is called an equivalent stress, the value of which is compared to the yield strength of the material to evaluate chances for irreversible deformation. Another failure mechanism is stress by material fatigue, which occurs in some materials as metals if they are loaded dynamically for long periods of time. The maximal stress that such a material can stand in cyclic loading indefinitely is called fatigue strength. Biological materials as bone may fail in addition through stress-related necrosis and resorption. The mechanisms for these biological failure modes are not well understood, however, and a stress criterion is not available as yet.[2] Nevertheless, these failure modes are very important in long-term performance of orthopedic and dental implants.

IV. STRUCTURAL STRESS ANALYSIS

A. Introduction

A structure is a geometric configuration of materials created to withstand loads. Structural analyses are performed to predict stresses occurring within the loaded structure, with the objective to evaluate the adequacy of its design and materials in fulfilling its load-bearing functions. The stress distribution in a structure depends on four, and only four, aspects:

1. The magnitude and configuration of the external loads, the loading conditions
2. The geometry of the structure
3. The (elastic) material properties
4. The physical nature of connections with the environment (boundary conditions) and between different materials (interface conditions)

For a structural stress analysis, these aspects must be described, either mathematically (in a theoretical stress analysis) or physically (in an experimental stress analysis). This description is called a model of the structure. The development of the model is probably the most difficult and crucial step in the analysis.

In the beginning of this chapter we have discussed that materials are assumed continuous, so that continuum theory can be applied. This was actually an important step in the process of modeling, enabling us to define and use the concepts of stress and strain as representing internal load and deformation on a local basis. The development of free body diagrams, too, is actually a modeling process. The objective of modeling is to "catch" the structure in a formal concept that can be described and solved by available techniques. This is done by introducing simplifying assumptions with respect to the four structural aspects mentioned above. The nature of these assumptions depends on what is known about these aspects, on the power of the methods of analysis, and on the objectives of the analysis.

We can roughly divide the objectives in three categories. In the first, design evaluation, are those analyses aimed at testing the mechanical performance of a specific design. A possible question, in this case, is simply whether the design (which may be an implant) will fail or not, based on quantitative criteria known. Another possibility in this category is to compare a specific design to another or previous one. The "model" in this case may be the structure itself, a prototype on which stresses are experimentally determined, simulating the loading environment for which it is intended. The second category, design optimization, is directed at investigating the effects of design parameters (geometry, material) on the stress distribution. Such an analysis may be aimed at a true mathematical optimization, e.g., to evaluate the one stem shape of a total hip replacement for which acrylic cement stresses are minimal, or it may be limited to investigate a few shapes to establish trends from which design decisions are made. For design optimization purposes, parametric analysis will usually be applied, investigating the isolated influences of structural properties on the stress distribution by subsequent variation of parameters. Theoretical methods are suitable for parametric analyses in particular. Obviously, the border line between the first and the second category is not very strict, nor is it between the second and the third, basic research. This category includes all analyses aimed at finding fundamental concepts with respect to the shape and function of the structure itself. Examples in this category are those pertaining to stress-related architecture and remodeling of bone; effects of material characteristics (such as anisotropy, viscoelasticity), geometry, and loading on stress distributions; development of special techniques for the analysis of biological tissues; principles of load transmission in bone-prosthesis structures; stress-related interface behavior; and so on. Parametric analysis is a major tool in this category.

The stress analysis itself is carried out either experimentally or theoretically. To avoid misunderstanding it must be remarked that experimental methods involve the application of mechanics theories as well, whereas theoretical stress analysis uses data that have been determined experimentally. Both methods are performed on models. If the analysis is experimental, a physical model is used, in a wide sense, e.g., including a prototype. In theoretical analyses, the model is mathematical, a set of equations. Theoretical methods can be divided into analytical and numerical ones. The first are used if the four structural aspects can be made to fit into a mechanics theory for which closed-form solutions exist. In this case, the mathematical equations are solved directly, yielding (closed-form) formulas in which the stress magnitudes are directly related to the parameters describing the structural aspects. An example of such a structure is the linear elastic bar in tension, discussed previously. Numerical methods (mostly finite element methods) are based on the use of computers. They yield their results in numerical form (based on numerical input data), so that no direct relation is obvious between stress values and structural parameters. These numerical methods are of unlimited applicability in principle, so very complex models can be solved.

When the results are obtained, their accuracy and validity must be examined. The first requires a sound understanding of the methods applied; the second requires a realistic assessment of the relation between model and reality. The last step is to determine the significance of the results: what do they mean for the problem at hand? To answer this question, the stresses predicted may be compared to stress failure criteria, based on specific failure mechanisms. These mechanisms, generally speaking, include fracture, irreversible deformation, and fatigue failure. Biological materials such as bone may ''fail'' also due to biological reactions to stress, by necrotizing, and resorbing.

The procedure of stress analysis can thus be divided into four steps: structure identification and assessment of objectives, model development (including the collection of required data), analysis process, result validation and interpretation. Several decisions have to be made in the execution of these steps, the consequences of which are closely interrelated. The central issue in the whole process is the model, the characteristics of which affect all steps.

B. Methods of Analysis
1. Experimental Methods

To understand the principles of experimental stress analysis, it must be appreciated that stresses cannot be determined in direct measurements. Just like heat becomes manifest by its effects only (temperature changes), stresses are abstract concepts that must be determined indirectly, by measuring their effects (i.e., deformations). Nor can strains be measured directly as they are (as stresses) defined in an infinitesimal small geometric point. Most experimental methods measure displacements between points, which are calculated to strain (in this case an average value over the measured region), and later to stresses (using the constitutive equations of the materials concerned). With few exceptions (e.g., 3-D photoelasticity and Moire fringe techniques), all methods can be applied to either a prototype or

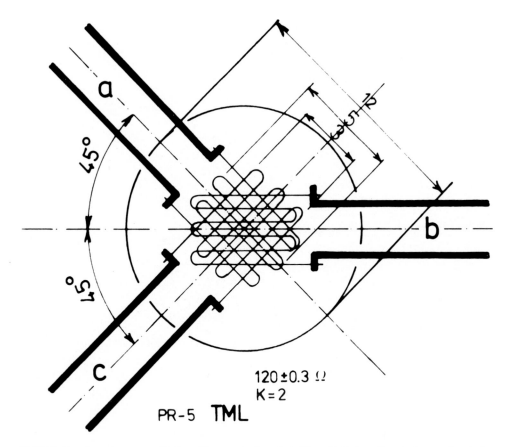

FIGURE 13.   A strain rosette with three filaments (a, b, and c). The coils that are the actual measuring devices are shown (measured region, 3 mm). The thick lines represent the connections to the measuring equipment.

a laboratory model of a structure. Again with the exception of 3-D photoelasticity, all methods are suitable for evaluating stresses at the free surfaces of structures only. In some methods, measuring devices are applied to the surface (e.g., extensometers, electrical strain gauges); in some the surfaces are applied with strain sensitive coatings (e.g., brittle or photoelastic coatings), while in others the displacements are registered directly with photographic or laser techniques (e.g., Moire, holography, interferometry). An excellent review of methods, their possibilities, and limitations was published by Durelli.[6] Here only two techniques frequently applied in biomechanics (hard-tissue) analysis, strain gauges and photoelasticity, are discussed.

Electrical strain gauges are based on the principle that the resistance of a material changes when stretched. The gauge consists of a small electrical coil that is glued on the surface of the structure, in a certain orientation. The change in voltage over the coil is directly proportional to the relative change in length of the coil. Thus, the strain registered is an average value over the length of the gauge in its specific direction. If a state of uniaxial stress (or uniaxial strain) occurs in the structure and the gauge orientation coincides with that of the stress, the measured strain (ϵ) can be calculated directly to stress (σ), using Young's modulus (E) of the material ($\sigma = E\epsilon$). In general, however, a state of plane stress presides at a structure's free surface. As discussed previously, the plane stress state is fully characterized by three independent variables (e.g., the values of two direct stresses and a shear stress or two principal stresses and a principal stress orientation). To completely determine all stress variables in this case, use can be made of a strain rosette, which essentially consists of three coils registering strains in three different orientations (see Figure 13). The three measured

FIGURE 14.   An embalmed human femur in a laboratory setting, applied with 100 strain rosettes on the diaphysis. (From Huiskes, R., Janssen, J. D., and Slooff, T. J., *Mechanical Properties of Bone*, Vol. 45, Cowin, S. C., Ed., American Society of Mechanical Engineers, New York, 1981, 211.)

values are calculated to principal strain magnitudes and orientation, and subsequently to principal stresses.

Obviously, stresses in structures can be calculated from strains only if the constitutive equations of the material are known. Because the strain gauge averages strain over a certain region (however small that may be), high stress gradients and local stress concentrations can usually not be determined adequately. In view of their heat sensitivity, temperature compensation during measurements is necessary. Strain gauges can also be applied for structures (or models) that are loaded dynamically. A potential disadvantage of this method is that many gauges are needed for a complete evaluation in the case that the stress distribution is irregular.

Strain gauges have frequently been used for stress analyses of bones, in vivo[7] and in vitro.[3] Figure 14 shows a cadaveric femur applied with 100 rosette strain gauges.[3] Strains were registered for different kinds of loads applied to the femoral head and calculated to principal stresses, using an assumed Young's modulus and Poisson's ratio of the bone material. An example of stress values, as distributed over the femoral shaft for an axial compressive load, is shown in Figure 15. Also shown in this figure are the stresses as calculated using a closed-form theory (beam theory will be discussed later). Although the experimental and theoretical results compare reasonably well, local differences are apparent.

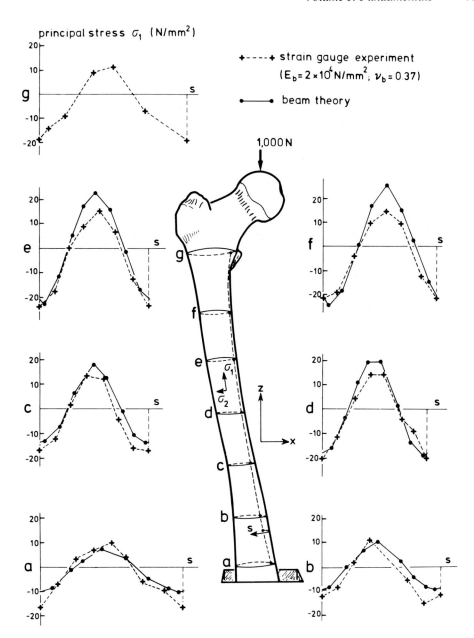

FIGURE 15.   Maximal principal stress (σ_1) on the femoral surface at seven levels as occurring around the periphery of the bone. Results are shown as calculated from the strain gauge experimental data (see Figure 14) and as calculated from a theoretical model. (From Huiskes, R., Janssen, J. D., and Slooff, T. J., *Mechanical Properties of Bone*, Cowin, S. C., Ed., American Society of Mechanical Engineers, New York, 1981, 211.)

These are a result partly of simplifying assumptions on which the mathematical model is based, but partly also of the assumed Young's modulus used to calculate the measured strains to stresses; anisotropy and local nonhomogeneity of the material in particular play a role here. Comparisons of this kind, between experimental and theoretical results, are important to establish the validity of models, as in this example the validity of a linear elastic, isotropic, and homogeneous representation of the bone can be assessed.

The limitations of the experimental approach are evident in this example. To obtain a complete characterization of the stress distribution, many rosettes (100 in this case) have to be applied. Gluing these to the bone surface and connecting them to the electronic measuring equipment is a time-consuming endeavor, which may take more than half a year's work. Moreover, the equipment needed is expensive, since so much measuring data must be collected automatically. In view of the time periods needed for preparation and experiment, the bone has to be embalmed with formalin, which affects its material properties. Hence, the experimental bone is merely a rough physical model of the original living bone. Examining the influences of load changes on the stresses is easily done, but variation of the material properties or the geometry of the model is impossible short of starting a new experiment. In addition, stresses can be determined on the outside surface of the bone only, whereas no information is obtained from within the material. As we will see in the next section, a theoretical technique as the finite element method is more suitable for a complete stress evaluation of irregular structures. Nevertheless, experiments are necessary to provide reference data for the development of mathematical models, whereas some problems are more suitable for an experimental approach.

Another technique which has been used frequently in hard-tissue biomechanics is that of photoelastic analysis.[8] This method uses the principle that deformation patterns become visible as fringes in certain plastics when illuminated by a polarized light source. The difference between indexes of refraction are proportional to the maximal shear stresses in the material. A disadvantage of the method (if not applied as a coating) is that it can only be used in physical models, to be made out of the photoelastic material with given (unchangeable) mechanical properties. A photoelastic model can be 2-D, in which case the fringes are directly visible and are quantified by photographic means, or 3-D, in which case the fringes have to be "frozen" at high temperatures and the model cut to evaluate the results. Contrary to other methods of experimental stress analysis, this technique also provides information inside the material of the model. Nevertheless, it has almost completely been replaced by finite element methods. In 2-D analysis, the method still has a use for instructional purposes since stress patterns can be made readily visible in simple configurations. In addition, photoelastic methods can be useful to locate stress concentrations and to evaluate the influences of complex boundary conditions that are difficult to model mathematically. An example is presented in Figure 16. The structure shown is a simple (2-D) model, representing the general characteristics of an implant, fixated in the medullary canal of a bone with acrylic cement. The cement layer is represented in the model by layers of photoelastic material. The aluminum structure is U-shaped to take the structural integrity between upper and lower "bone" slabs into account. A transverse load is applied and the photograph shows the resulting fringe patterns in the photoelastic material. Each fringe represents one refraction index increase in the maximal shear stress. Apparently, stress concentrations occur in a relatively small region of the "cement" (the outer, "proximal" part). A similar high-stressed region, not shown in this figure, occurs on the other ("distal") side. Hence, most of the load transference between "stem" and "bone" takes place in these two locations. Secondly, it appears that the stress distribution close to the boundaries of the photoelastic slab is very sensitive to the surface finish (the boundary conditions of the photoelastic material). Figure 17 shows a result of a finite element analysis of the same structure.[10] The proximal and distal high-stressed regions are apparent in these results too. However, the influences of irregular interface contact are not seen in this case because they are assumed smooth in the theoretical model.

2. Closed-Form Theories

In the history of solid mechanics closed-form theories have been developed, formulated, and verified for a great number of problems. The essence of closed-form solutions is that stress and strain values are expressed directly in the parameters describing the structural

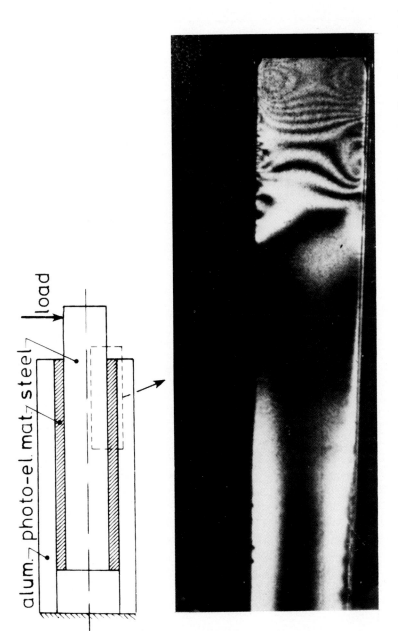

FIGURE 16. Local photoelastic fringes in a general model of an intramedullary fixation system. The fringes represent lines of equal maximal shear stress in the photoelastic material that represents acrylic cement. (From Crippen, T. E. and Huiskes, R., unpublished data, 1980.)

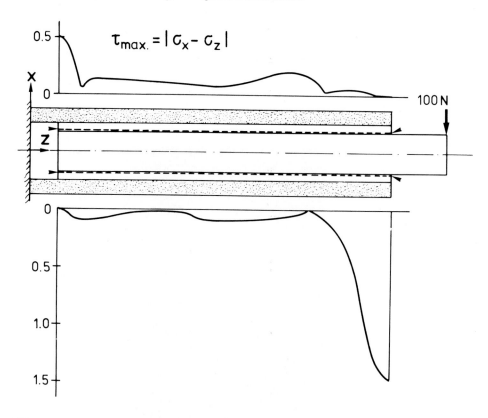

FIGURE 17.   Maximal shear stresses in acrylic cement near the cement-stem interface in a general model of an intramedullary fixation system on either side of the stem (compare Figure 16). The stresses were calculated by 2-D finite element method.[10] The cement-stem interface is assumed loose in this particular example, comparable to the experimental model of Figure 16. Stress concentrations at the "proximal" and "distal" sides are evident. Because $\tau_{xy} = 0$ at the (sliding) stem-cement interface, the maximal shear stress in the cement material at this interface $\tau_{max} = |\sigma_x - \sigma_z|$.

aspects. Examples of closed-form theories that are frequently applied in biomechanics are linear-elastic bar theory, beam theory, and torsion shaft theory. All three are valid for slender, prismatic bodies whose length is much larger than the thickness, made out of linear-elastic, isotropic, and homogeneous materials, loaded in axial tension or compression, transverse forces or bending, and torsion, respectively. These theories are available for 2-D and 3-D structures, uniform and variable cross sections, and straight and curved bodies. Although, particularly in biomechanics, structures seldom behave in accordance with these theories exactly, approximate models can sometimes be developed that do. Prosthetic stems, bone pins, and rods are good examples of structures that yield to these theories, as well as bone plates (although the holes present a complication), and diaphyses of long bones by rough approximation. Other simple closed-form theories used in hard-tissue biomechanics are compound-beam theory (applied to bone-fixation plate and bone-prosthetic stem composites), beam-on-elastic-foundation theory (prosthetic stem and tibial plateau fixation), and plate and shell theories (skull).

In many textbooks on biomechanics, only the closed-form theories for relatively simple structures are treated. One must realize, however, that in biomechanics specifically, regular structures are exceptions rather than rules. Although the simple theories are more easily explained, require less complicated mathematics, and thus are attractive for teaching purposes, it must be appreciated that this material is of limited applicability.

Closed-form solutions are attractive for obtaining rapid, approximative solutions for certain

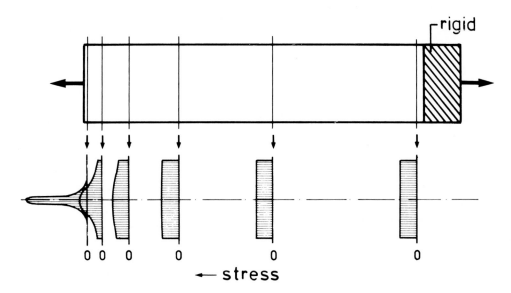

FIGURE 18. The force that loads the bar on a small area (left) engenders end effects in the bar: the stress distribution in the bar cross sections (shown below) becomes uniform only after a certain length. The rigid block on the right side transmits a uniform stress distribution immediately. Due to requirements of equilibrium, the average stress must be the same in all sections (σ_{av} = F/A).

problems. They often give a direct insight into relations between essential structural parameters and stress behavior. Their application is cheap and rapidly leads to results, and they are therefore often applied to obtain first-order reference solutions for more advanced experimental and numerical analyses. Many of the closed-form theories traditionally applied, however, in particular those that are mathematically complex, have become obscure in the recent past because of the much greater potential of finite element analysis.

a. Bar Theory

We will limit ourselves to straight bars of uniform cross section, as discussed previously (see Figure 11). The essential feature of bar theory is that a plane cross section remains plane after the load is applied and parallel to the plane as it was before deformation. As a consequence, the stress state is uniaxial, uniformly distributed over a cross section and throughout the bar. We have seen that the stress can be calculated directly from the force and the cross-sectional area (σ = F/A) and the strain can be calculated from the stress and Young's modulus (ϵ = σ/E). From these two equations, it follows immediately that ϵ = F/EA. The entity EA thus governs the deformation of the bar and is called the axial rigidity, a structural parameter combining material and geometrical properties.

In order for the bar cross sections to deform parallel to their original shape, the force must apply in the center of gravity of the section. Close to the point of force application, the stress distribution is disturbed by end effects (Figure 18, left side): some length is needed for the stress distribution to become uniform. On the right side of Figure 18 the force is introduced through a rigid block, which immediately provides for a uniform distribution. Comparing both sides of Figure 18, it appears that if far enough removed from the region of load application, the nature of the load application becomes irrelevant. This is a general rule, known as the principle of Saint-Venant. It implies that in any stress analysis, local disturbances in the stress distribution (i.e., stress concentrations) must be expected in the neighborhood of applied loads, which are often not predicted in the results. Hence, the region of interest must either be far enough removed from the point of load application or the boundary conditions must be accurately reproduced. In bars as used for materials testing

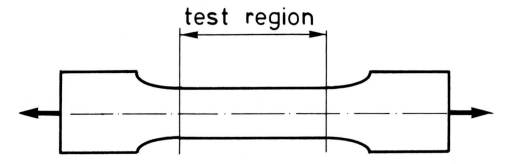

FIGURE 19. Samples as used for materials testing are shaped in a particular form to guarantee a uniform stress distribution in the test region. End effects are eliminated in this way.

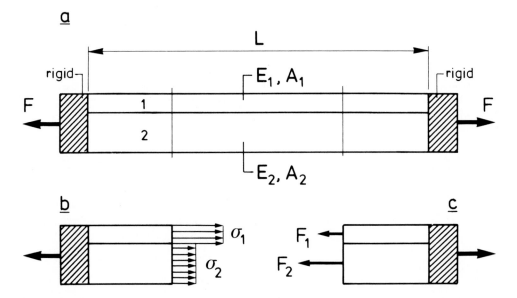

FIGURE 20. (a) A compound bar (a combination of the Bars 1 and 2) loaded in tension. The strain in the bar is uniform owing to the rigid blocks on the sides. (b) Due to differences in cross-sectional areas (A_1 and A_2) and Young's moduli (E_1 and E_2) of the individual bars, the stresses differ. Evidently, (c) each bar carries a part of the load.

(Figure 19), the influences of end effects are eliminated by a particular shape that guarantees a uniform stress distribution in the test region.

b. Compound-Bar Theory

Assume a bar as shown in Figure 20a, a combination of two bars with different material properties, loaded in tension. End effects are eliminated by rigid blocks at the ends. In compound-bar theory, it is assumed that the cross sections of the bar as a whole remain plane and parallel after the load is applied, a condition imposed here through the rigid blocks. This condition implies that the strain (ϵ) is uniformly distributed over a cross section. Therefore, the stress by necessity is not. In material 1, we find by applying the constitutive equation: $\sigma_1 = E_1\epsilon$, and in material 2: $\sigma_2 = E_2\epsilon$; since E_1 and E_2 differ, σ_1 and σ_2 cannot be equal either (see Figure 20b). We know that the total internal force in a cross section must be F, but how large are the internal forces in each material in each individual bar (see Figure 20c)? Obviously $F_1 = A_1\sigma_1$ and $F_2 = A_2\sigma_2$. Let us assume that the load (F), the material properties (E_1 and E_2), and the geometry (A_1 and A_2) of this structure are given, and we wish to evaluate the stresses (σ_1 and σ_2). We have five equations at our disposal,

Equilibrium: $\quad F_1 + F_2 = F \qquad\qquad\qquad (1)$

Constitutive eq.: $\sigma_1 = E_1\epsilon \qquad\qquad\qquad (2)$

$\qquad\qquad$ and $\sigma_2 = E_2\epsilon \qquad\qquad\qquad (3)$

Bar theory: $\quad F_1 = A_1\sigma_1 \qquad\qquad\qquad (4)$

$\qquad\qquad$ and $F_2 = A_2\sigma_2 \qquad\qquad\qquad (5)$

and five unknowns (F_1, F_2, σ_1, σ_2, and ϵ). Hence, the unknowns can be evaluated. From Equations 2 and 4 and 3 and 5 we find:

$$F_1 = A_1 E_1 \epsilon \quad \text{and} \quad F_2 = A_2 E_2 \epsilon$$

hence,

$$\frac{F_1}{F_2} = \frac{A_1 E_1}{A_2 E_2}$$

combining this with Equation 1, it follows:

$$F_1 = \frac{A_1 E_1}{A_1 E_1 + A_2 E_2} F \qquad\qquad F_2 = \frac{A_2 E_2}{A_1 E_1 + A_2 E_2} F$$

and applying Equations 4 and 5 again,

$$\sigma_1 = \frac{E_1}{A_1 E_1 + A_2 E_2} F \qquad\qquad \sigma_2 = \frac{E_2}{A_1 E_1 + A_2 E_2} F$$

Apparently, the internal load in each individual bar depends on the ratio between its own axial rigidity and the axial rigidity of the compound bar. This means that when bar 1 is flexible and bar 2 is stiff ($E_1 A_1 \ll E_2 A_2$), most of the load will be carried by bar 2, and vice versa.

It can be proven that there is no shear stress at the connection between the two bars in the above case. Hence, whether they are connected or not is irrelevant. This situation changes when only one of the bars is loaded at the end (Figure 21). Then an end effect occurs, which is a result of load transmission from bar 1 to bar 2. We will not go into the mathematics of this phenomenon, but discuss it qualitatively. At the right-hand side, all load is applied to bar 1. Then gradually it is transferred to bar 2 by shear stresses at the connection, the interface, between the two bars. After a specific length the end effect vanishes and the internal loads in the two bars are divided as calculated above. In other words, from there on the structure again acts according to compound bar theory. The consequences of this load transfer mechanism for the internal loads in the two bars separately and the shear stresses at the interface are shown schematically in Figure 22. The length of the load-transmission region depends on the structural properties of both bars.

Let us now assume that the forces are applied to each bar separately (see Figure 23a). In that case the end effects occur on each side. When the fixation length (L) is decreased, the load transmission regions will eventually meet. From a certain length (see Figure 23b), the interface shear stress does not vanish anymore between the ends, but has a more or less uniform value over the full fixation length. In that case, the shear stress approximately equals the force divided by the fixation area ($\tau = F/d\ell$, where d is the thickness of the bars). What is to be learned from this example is that one has to be careful with the notion

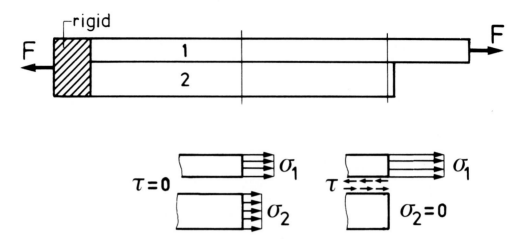

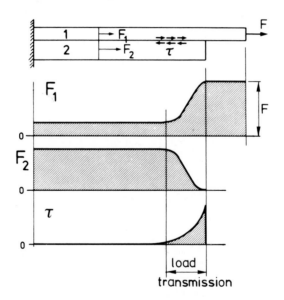

FIGURE 21.   A compound bar as in Figure 20, but this time the load applies to Bar 1, as a consequence of which load transmission takes place through shear stresses (τ) from Bar 1 to Bar 2. This "end effect" (shear lag) vanishes after a certain length, and from there on the compound bar behaves in accordance with compound bar theory.

FIGURE 22.   Internal load distribution in the compound bar of Figure 21. Internal forces in the separate bars (F_1 and F_2) and the interface shear stress are shown along the length. The load transmission phenomenon (shear lag) is evident.

that stress "is force divided by area". This is only true on a local basis. It is not universally so that when "the area is enlarged the stress decreases". Enlarging the fixation length of the bars in Figure 23a has no effect on the magnitude of the shear stresses; the load transmission regions only separate further in that case.

c. Beam Theory

A beam is a slender, prismatic body loaded in pure bending or bending by transverse forces. We will limit ourselves to straight beams with uniform, symmetric cross sections, loaded in one plane and made out of linear elastic, isotropic, and homogeneous materials. The essential feature of beam theory is that plane cross sections remain plane (but not

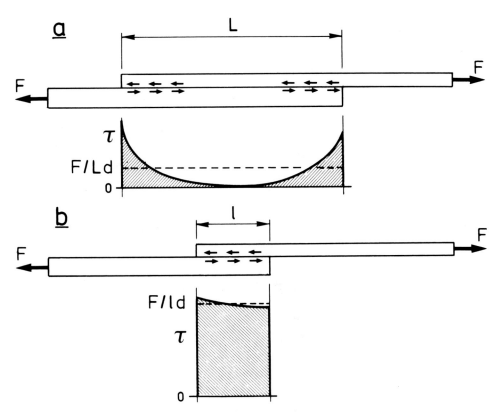

FIGURE 23.   A compound bar as in Figures 20 and 21, but this time the loads apply to each bar separately on either side. Load transmission (as in Figure 22) occurs (a) on each side and vanishes toward the middle region of the fixation, as witnessed by the shear stress distribution. (b) If the fixation length decreases, the load transmission regions meet and the shear stress becomes more or less uniform. In both cases the average shear stress equals force divided by area (F/Ld and F/ℓd, respectively). However, elongating the fixation in the first example (a) only serves to separate the load transmission regions further, whereas the peak stresses will not decrease.

parallel) after the load is applied. As in bar theory, the stress state is uniaxial, but the stresses are not uniformly distributed over a cross section. Let us first consider the case of pure bending by a moment M (Figure 24). Equilibrium conditions dictate that the only nonzero internal load in an arbitrary cross section is a bending moment (M). Thus, the stress distribution in the cross section is such as to have a zero axial force and a nonzero bending moment as resultants. Furthermore, the stress distribution must be linear because the strain distribution is linear in view of the fact that plane cross sections remain plane after deformation. These requirements can only be fulfilled by a stress distribution as shown in Figure 24. The stress is zero in one point. The collection of these points in all cross sections of the beam is called the neutral line (or rather neutral plane, taking the third dimension into account). The stresses are negative (compression) on one side of this line and positive (tension) on the other. It can be shown that the stress value at a given point with distance y from the neutral line:

$$\sigma = My/I$$

in which I is the static (or second) moment of inertia (SMI) of the cross section in the plane considered. I depends on geometrical parameters only. In the case of a rectangular cross section, as in the example of Figure 24:

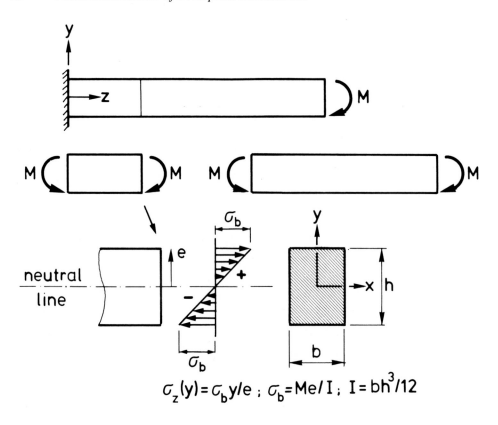

$$\sigma_z(y) = \sigma_b y/e \; ; \; \sigma_b = Me/I \; ; \; I = bh^3/12$$

FIGURE 24.   A cantilever beam loaded in pure bending. The stress distribution in a cross section is nonuniform, but linear. Stresses are zero in the neutral line. The bending resistance of the beam depends on the second moment of inertia (I), which depends on the cross-sectional geometry.

$$I = bh^3/12$$

The maximal stress value in the section (in absolute value equal on either side because the section is symmetric) is often called the bending stress (σ_b). Obviously,

$$\sigma_b = Me/I$$

in which e is the maximal value of y (in the example e $=$ h/2).

The SMI plays an important role in bending of beams, comparable to that of the cross-sectional area in axial loading. It can be shown that the angular deflection ϕ^* of a cross section after load application in bending equals:

$$\phi = M/EI$$

where E is the Young's modulus of the material. Thus, the deflection is governed by the value EI, a structural parameter called the flexural rigidity of the beam. It plays a role in beam theory and compound beam theory comparable to that of the axial rigidity (EA) in bar theory.

Let us examine the SMI a little closer. In the example of Figure 24, the bending takes place in the y-z plane, around the x axis. Hence, the relevant SMI in this case is the one around this axis ($I_x = bh^3/12$). If bending would take place in the x-z plane, then the relevant

---

* ϕ must be expressed in radians.

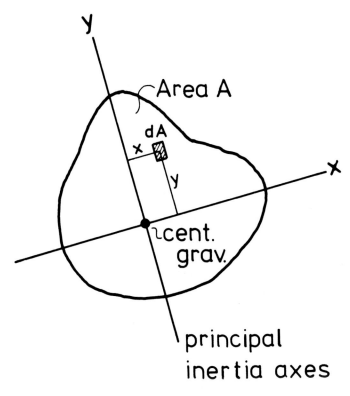

FIGURE 25.   An arbitrary cross-sectional shape can be characterized by two principal inertia axes. These axes cross the area center of gravity and have the property that if bending occurs around either of these axes, the axis concerned is incorporated in the neutral plane. The moments of inertia are calculated by integrating the contributions of each infinitesimal area part, which is the product of its area (dA) and its squared distance to the axis concerned.

SMI would be the one around the y axis ($I_y = hb^3/12$). Generally speaking, a cross-sectional area has two principal axes of inertia (Figure 25). If bending takes place around one of these axes, it forms a part of the neutral plane. In that case, the formula for the stress distribution given above is valid. The principal inertia axes cross in the center of gravity of the section area. Every small material part in the cross section contributes to the SMIs by a magnitude equal to its area (dA) multiplied by its squared distance to the axis concerned. Integrating these contributions over the total area (A) gives (see Figure 25):

$$I_x = \int_A y^2 dA \qquad \text{and} \qquad I_y = \int_A x^2 dA$$

The largest value of these two is called the maximal principal SMI (I_{max}) and the smallest is called the minimal principal SMI (I_{min}).

In section areas that are symmetric, the inertia axes coincide with the symmetry axes, as in the example of Figure 24. Figure 26 shows examples of SMI formulas for a few regular shapes. Figure 27 shows a number of shapes all having the same area magnitude, but different SMI values. As evident from this comparison, the material away from the axis contributes most to the SMI, as can also be seen in the general formula. Apparently, the hollow circular shape of long bones offers good bending resistance in all directions.

If a beam is loaded by a transverse force (Figure 28), the nonzero internal loads in a cross section are a bending moment (M) and a transverse force. The bending moment in this case

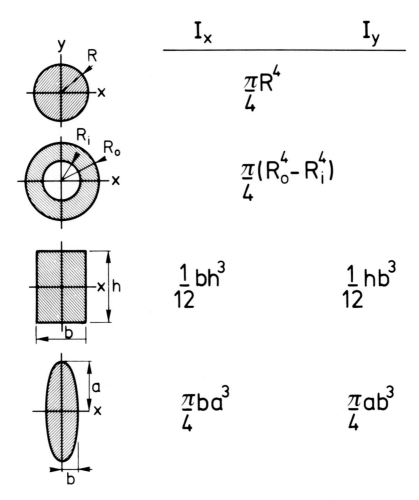

FIGURE 26.   Principal moments of inertia of a few shapes. All shapes are symmetric, hence the positions of the principal inertia axes are obvious.

is not distributed uniformly throughout the beam, but depends on the distance to the force line of application: $M = F\ell$. The internal transverse force at all sections equals the external force F. The internal bending moment M represents an axial stress distribution completely equal to that discussed previously for the case of pure bending. The internal transverse force represents a shear stress distribution that is parabolic.

d. Torsion of Circular Shafts

We assume a straight long and slender shaft of uniform axisymmetric (circular) cross section, loaded in torsion and made out of a linear elastic, isotropic, and homogeneous material (Figure 29). In simple torsion theory (which applies to this structure), it is assumed, again, that plane cross sections remain plane after application of the load. The torque (M_t) results in an angular twisting of the shaft around the longitudinal axis. The only internal load in the cross sections, uniformly distributed along the shaft, is the moment M_t, which represents a shear stress distribution in the plane of the section; all other stress components in the cross section are zero.* The shear stress increases from zero at the center to maximal at the periphery and is calculated from

---

*   Torsion does produce direct stress working in other sections, not perpendicular to the shaft axis.

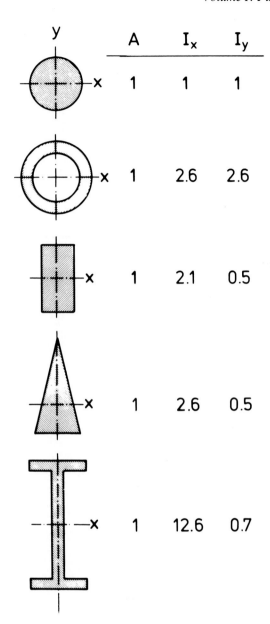

FIGURE 27.   Examples of a few cross-sectional shapes, drawn on the same scale, that all have equal areas, but vastly different moments of inertia. Evidently, material on the periphery of the sections contributes most. If bending can occur around an arbitrary axis, then the hollow circular shape as it occurs in bones gives optimal resistance for minimal use of material.

$$\tau = M_t r / I_p$$

in which r is the radius to the point concerned, and I_p is the polar moment of inertia.

In general $I_p = I_{max} + I_{min}$, hence for a massive circular cross section (compare Figure 26),

$$I_p = \frac{\pi}{2} R^4$$

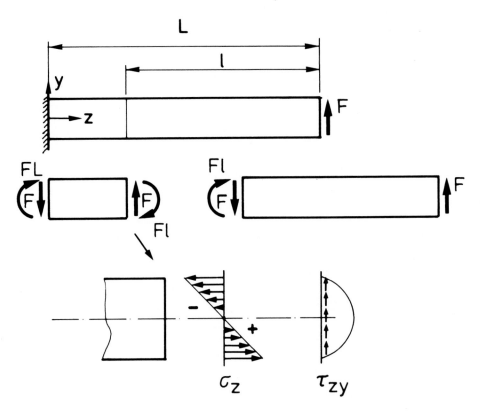

FIGURE 28.   A cantilever beam loaded by a transverse force engenders a bending moment in the beam cross sections (as in Figure 24) plus a transverse force. The first can be treated as in pure bending; the latter is the resultant of a parabolic shear stress distribution. Other than in the case of pure bending, the internal bending moment changes along the length of the beam (compare Figure 5).

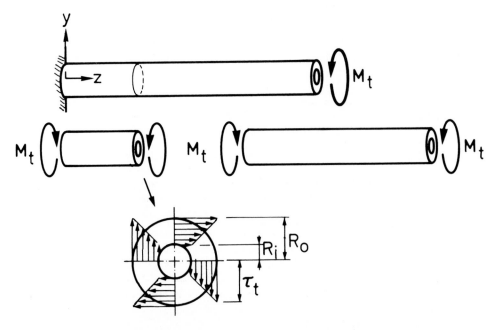

FIGURE 29.   A circular, hollow shaft loaded in torsion. In each cross section works a torsional moment which is the resultant of a shear stress distribution in the plane of the section, as shown. The magnitude of the stress is proportional to the distance to the central axis.

and for a hollow tube,

$$I_p = \frac{\pi}{2} (R_o^4 - R_i^4)$$

where R_o is the outer and R_i is the inner radius. Obviously, the maximal shear stress in a cross section is

$$\tau_t = M_t R_o / I_p$$

If θ is denoted as the angular twist per unit length* of the shaft, it can be shown that

$$\theta = M_t / G I_p$$

in which G is a material parameter called the shear modulus, to be calculated from Young's modulus and Poisson's ratio according to $G = E/2 (1 + \nu)$. In analogy to beam and bar theory, the value GI_p, which governs the angular twist, is called the torsional rigidity of the shaft. It is evident from the formulas for I_p (compare also Figure 26) that material on the periphery of the section contributes most to the structural rigidity, comparable to bending. As follows from Figure 27, a hollow shaft is able to produce significantly more torsional rigidity than a massive one if the amounts of material (the cross-sectional areas) are equal. It must be remarked, however, that when the integrity of the cross section is disturbed, for instance, by holes or cracks, so that it is no longer closed, the torsional rigidity reduces dramatically.

It is always tempting to use simple closed-form solutions for structures more complex than the ones they were intended for. In other words, to simplify a model to such an extent that a simple theory can be applied. Although this is not bad in principle and, depending on the objectives, may even represent good engineering practice, it is obvious that the consequences of the simplifying assumptions for the validity of the model should be well understood. Simple torsion theory in particular is rather unforgiving where its underlying assumptions are concerned. Shafts that are not circular in cross section do not yield to the theory, but can be analyzed with Saint-Venant's warping theory,[24] which is many times more complex. An example of a shear stress distribution in a femoral cross section loaded by torsion, determined with this theory, is shown in Figure 30. For the solution, a numerical process was used. It is evident from these results that although the shape of the section does not deviate dramatically from circular, the shear stresses are not just proportional to the radius. A more precise characterization is that they depend on the width of the cortical shell, being maximal where the section is narrow, somewhat comparable to the fluid velocity in a stream through a river bed.

e. Combined Loading of Slender Bodies

Loading of slender bodies (as bars and beams) by axial forces, transverse forces, and bending moments results in axial direct stresses, normal to the cross sections. Although transverse forces generate shear stresses as well, the axial direct stresses are usually the most significant ones where chances for failure are concerned. If these loading cases occur in combination, then the stresses in the structure are found by superimposing the stresses as they would result from each loading case separately. This is an important principle in stress analysis, valid fortures in general as long as they behave linear elastic.

Assume a femur as shown in Figure 3, loaded by a force on the head and cut (freed) at

---

* Note that the unit of θ is radians per unit of length (mm^{-1}).

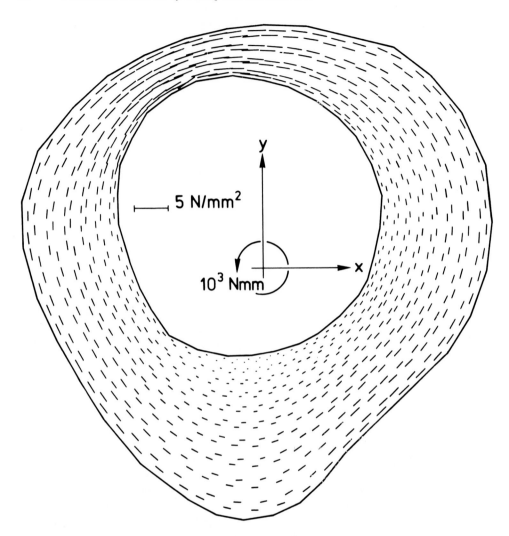

FIGURE 30.   Shear stress distribution of a (noncircular) section of a femur loaded by a torsional moment, calculated with Saint-Venant's warping theory, using numerical methods. The orientations and lengths of the line pieces represent directions and magnitudes of the stresses. Evidently, the distribution differs from that in Figure 29; maximal stresses occur specifically there where the cortical shell is narrow. (From Huiskes, R., Janssen, J. D., and Slooff, T. J., *Mechanical Properties of Bone*, Vol. 45, Cowin, S. C., Ed., American Society of Mechanical Engineers, New York, 1981, 211.)

a particular cross section. Using the principles described previously, the internal (3-D) loading configuration in that section is determined. Generally, it consists of three moments (two bending moments and a torque), an axial force and two transverse forces (see Figure 31a). Only the bending moments and the axial force produce axial direct stresses, in which we are interested. We assume the femoral shaft to behave as a linear elastic, isotropic, and homogeneous 3-D beam. In other words, we treat the cross section as if it were a part of a bar/beam as we have discussed previously, the only difference being that the cross sectional shape is irregular (see Figure 31b). To apply the formulas of bar/beam theory, we have to evaluate:

1.   The cross sectional area (A)
2.   The center of gravity location

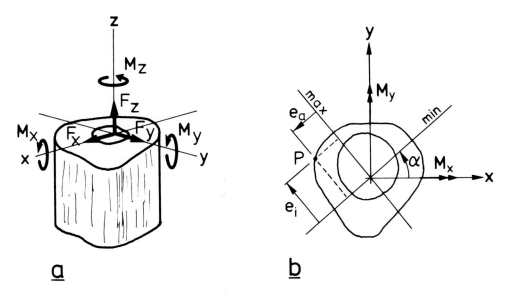

FIGURE 31.   (a) 3-D internal loads in a cross section of a femoral diaphysis. The axial force and the bending moments produce axial direct stresses in the cross section. (b) The orientations of the bending moments with respect to the principal inertia axes are shown. The axial direct stress in an arbitrarily chosen point P can be calculated by superimposing the separate contributions of the three loads.

3.   The principal inertia axes' orientations (α) with respect to the global (x-y) coordinate system
4.   The maximal and minimal SMI values (I_{max} and I_{min})

The evaluation of these parameters is done by digitizing the cross-sectional shape and applying the appropriate formulas.

   The axial direct stresses (σ_z) are then calculated by superimposing the individual contributions of the axial force and the two bending moments, applying the formulas discussed earlier. Assume an arbitrary point P in the section (see Figure 31b). The stress in that point resulting from the axial force (in fact in the whole section, since the stresses as resulting from this force are uniformly distributed) is

$$\sigma_z = -F_z/A \text{ (compression is negative)}$$

To calculate the stresses resulting from the bending moments M_x and M_y, these loads must be decomposed in components around the principal inertia axes:*

$$M_{max} = M_y \cos \alpha - M_x \sin \alpha$$
$$M_{min} = M_y \sin \alpha + M_x \cos \alpha$$

For the contribution to the stress at point P we find:

$$\sigma_z = \frac{M_{max} e_a}{I_{max}} + \frac{M_{min} e_i}{I_{min}}$$

Superimposing the contributions of the axial force and the bending moments gives:

---

*   Moments are treated as vectors here. The moment vector is identified by a double arrow point; the moment orientation is related to the vector by the right-hand rule.

$$\sigma_z = -\frac{F_z}{A} + \frac{e_a}{I_{max}} (M_y \cos \alpha - M_x \sin \alpha) + \frac{e_i}{I_{min}} (M_y \sin \alpha + M_x \cos \alpha)$$

Since all loads (F_z, M_x, M_y) and structural parameters (A, I_{max}, I_{min}, α) are known, σ_z can be calculated for all points in the section, substituting the appropriate values of e_a and e_i. An example was shown in Figure 15 (limited to points at the periphery, but in several cross sections) and compared with experimental data. As we have seen, the comparison is reasonably good, but deviates at some locations, due to simplifying assumptions in the theoretical model, but also in the analysis of the experimental data.

It is not the aim of this exercise that the reader can reproduce it in detail. It serves handsomely to illustrate 3-D beam theory and the execution of theoretical analysis in general:

- Identify the structure (the bone, freed from its environment) and assess the objectives (evaluate axial direct stresses in cross sections to judge the applicability of 3-D beam theory).
- Develop the model (free body diagram, cutting, assumptions of constitutive equations, bar/beam theory, principle of superposition), including the evaluation of required data (area, moments of inertia, loading characteristics).
- Process the analysis (combine the equations to a mathematical solution, calculate stresses in all relevant points).
- Result validation (by comparing with experimental data) and interpretation (yes, the femoral diaphysis does by a certain degree of approximation behave as a 3-D linear elastic, isotropic, and homogeneous beam).

Two final remarks have to be made. First of all, the validation of a model by experimental data is not always necessary. For instance, in any new problem related to this one, we can use the knowledge obtained here (or elsewhere in the literature). Secondly, if the result was not acceptable, model refinement would have to be carried out by restarting the modeling cycle, comparable to what was discussed with respect to torsion of bone diaphyses.

3. Finite Element Methods

The use of closed-form theories is limited to structures of relatively simple geometrical and material properties. Only for a very few types of regular structures (e.g., bars, beams, plates, and shells) do these theories yield relatively simple formulas which can be readily solved by hand. The finite element method (FEM) is essentially a computer method.[11] It can be used in principle to calculate stresses in load carrying structures of unlimited complexity, although there are limitations of a practical nature. Nevertheless, the method is preeminently suited for stress analyses of irregular structures as bones and bone-prosthesis structures and is meeting increasing interest in biomechanics research.[12,13]

A FEM model describes the four relevant aspects of a structure (loading conditions, geometry, material properties, boundary/interface conditions) in discrete, numerical form. The central issue is the geometrical description by elements: the structure (or rather, the model) is mathematically divided into a set of connected blocks, or elements, which can have various shapes (Figure 32). At their corners, faces, or edges, these elements have nodal points by which they are considered to be attached to each other. The type of element used depends on whether a model is 2-D or 3-D and on specific requirements of accuracy.

A few element shapes are shown in Figure 32: a 3-D (20-node) hexahedron element (a), suitable for arbitrary 3-D geometries (the elements can be "molded" in any shape, provided they are not overstretched or deteriorated, in one direction), a (quasi) 3-D (6-node) ring element (b) for axisymmetric structures only, two 2-D triangular (c and d), and a 2-D quadrilateral element (e).

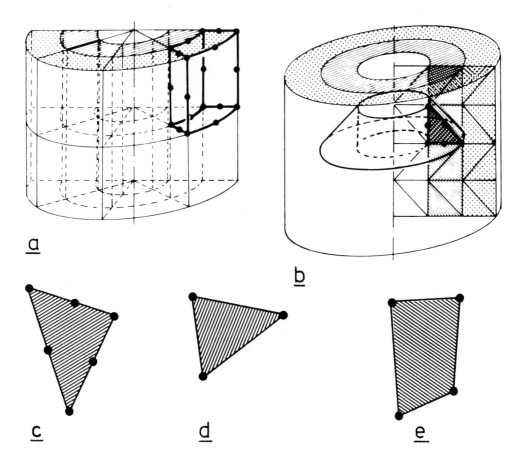

FIGURE 32. Examples of element types used in finite element analysis. (a) 3-D; (b) quasi-3-D; and (c, d, e) 2-D elements. Nodal points are indicated. (From Huiskes, R. and Chao, E. Y., *J. Biomech.*, 16, 385, 1983. With permission.)

Figure 33 shows the element mesh of a model. The structure concerned is part of a finger joint prosthesis, of which the metal stem is fixated within the medullary canal with a plastic plug. The model is axisymmetric and ring elements are used (see Figure 32b). The element mesh is usually generated automatically (at least partly) by a computer program, called a FEM preprocessor, based on varying amounts of data to be prepared by hand. This program organizes the mesh numbering, generates a nodal-point location file (the element numbers and accessory nodal point numbers), and a nodal-point coordinate file (the nodal-point numbers and their coordinates in an external reference system). In these two files the element mesh is completely characterized. The FEM program further requires an element characteristics file (for each element, its material properties, such as Young's modulus and Poisson's ratio in a linear elastic, isotropic problem, and in the case of a 2-D model, its thickness), a loading file, and a displacement file. The loading file contains the numbers of the nodal points at which forces are introduced to the structure (see Figure 33), the direction of these forces, and their magnitudes. The displacement file characterizes the kinematic boundary conditions. The model in Figure 33, for instance, is considered to be connected to the environment (supported, fixed) on the right-hand side, and the displacements of the nodal points concerned are prescribed as zero (suppressed). A nodal-point coupling file handles the characterization of the connections between nodes of different materials (regions or substructures). The stem in Figure 33, for example, is able to slide within the plug, for which the nodal points at the interface are disconnected in that direction (but not in transverse direction).

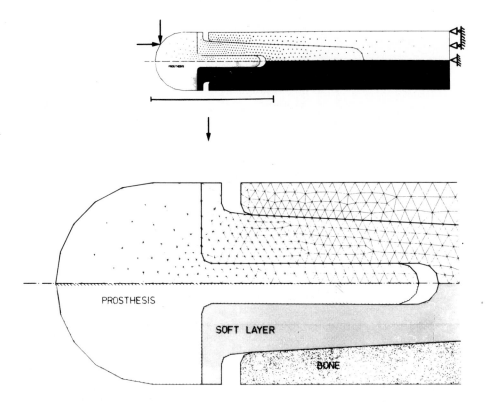

FIGURE 33. Element mesh in a model of a finger joint fixation system. The elements are axisymmetric (see Figure 32b), hence only a longitudinal section is shown. The loading is nonaxisymmetric.[14]

Based on this information, the FEM program calculates the displacements of all nodal points (which characterize the deformation of the loaded structure) and the stresses in all nodes or in the element centroids. The solution is obtained numerically through a set of linear equations, equal to the amount of degrees of freedom in the model: the number of nodal points times the number of displacement components in each node (two in a 2-D, three in a 3-D model). The computer time and memory space required for a problem progressively depend on the number of degrees of freedom. A time-efficient element mesh is of crucial importance, since computer capacity is the only practical limit to the level of model complexity feasible.

The FEM as applied in stress analyses is based on the principle of minimal potential energy, stating that a loaded structure deforms in such a way that the energy stored within it is minimal, a criterion which is formulated mathematically. The characteristic feature of the FEM is now that the total energy in the structure is discretized in individual contributions of each element. All (internal) forces working on an element are assumed to be concentrated in the nodal points. The relation between these nodal point forces and the nodal point displacements is expressed by a collection of parameters, called the element stiffness matrix. The parameters in this matrix depend on the specific nature of the element and on the properties of the material it represents. By applying the action = reaction law to nodal points (a node belongs to more than one element), requirements of compatibility (an element must not deform in such a way that it ''looses'' its neighbors), and the element stiffness matrices, the stiffness matrix of the whole structure is assembled. This matrix characterizes the energy in the structure and delivers the values needed to solve the linear equations mentioned above.

It will be clear from this brief description that the FEM is easier used than understood.

COARSE MESH

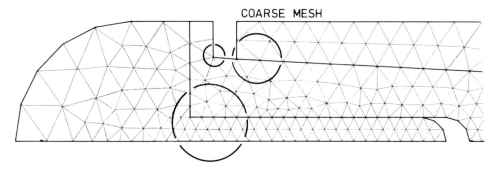

FIGURE 34. A coarse element mesh for the same structure as in Figure 33. In the circled areas, high stress gradients are expected. To represent these adequately, local mesh refinement is necessary.[14]

Once a suitable computer program is available, its application requires only some basic knowledge of computer communication and the method itself can be considered as a ''black box''. This feature of the FEM presents a great potential for misuse because it is commonly forgotten that the development of a FEM model which produces valid results, and a correct interpretation of these results, requires a basic understanding of the FEM principles. One essential issue to remember is that due to the discretization of the energy function and the assumptions needed to characterize the element load-deformation relations, the values obtained are approximations of the exact solutions. The approximations are such that they will converge to the exact solution when the number of elements, the mesh density, increases to infinite.* This implies in practice that in regions where high stress gradients (stress concentrations) are expected the mesh density must be relatively high. This is illustrated in Figure 34, showing a coarse mesh for the same structure as in Figure 33; the circled areas are those in which high stress gradients are expected, requiring (local) mesh refinement for adequate representation. To anticipate this correctly is purely a matter of experience, while testing these conditions objectively is easily done by repeating the calculations with a finer mesh; if the results are not significantly different after mesh refinement, the original mesh was adequate. This procedure is called a convergence test.

We have seen that the mesh density controls both the costs (computer time and space, which are limited) and the potential accuracy of a FEM model. To find the right balance by sophisticated modeling is the true skill of FEM analysis. Using meshes of relatively low density in those regions where stresses will not vary greatly or where results are not relevant is one way of reducing the costs. Reducing a problem from 3-D to 2-D, if such is possible, gives an enormous gain in efficiency. Modeling a 3-D structure as axisymmetric is another way of reducing complexity. Although a truly 3-D model has more potential where the validity of results are concerned, its practical potential for accuracy is less than that of a 2-D (or axisymmetric) model because of mesh density limitations dictated by computer capacity. FEM models usually have to be applied for parametric analysis, investigating the effects of structural properties and loading, for which several computer runs have to be made. The expenses for such analyses in the case of a complex 3-D model may become astronomical.** Another motivation for model simplicity is that the interpretation of complex 3-D model results is time consuming and difficult, while basic mechanisms are often equally well or even better studied with simpler models. A good practice is to apply complex models that closely mimic a structure to obtain reference data, while simpler FEM models (or closed-form theories) are used for parametric analysis.

---

* The exact solution is only ''exact'' with respect to the model applied, not to the (more complex) reality.

** If the loading conditions of a FEM model are changed, only a very small part of the FEM program has to be executed again. For all other changes, the program must be restarted. If the geometry is altered, the element mesh must be regenerated as well.

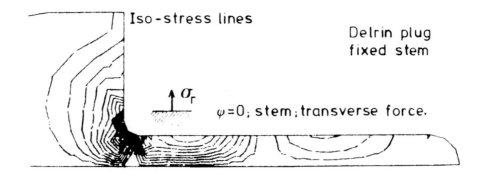

FIGURE 35.   Iso-stress lines in a section of the metal part of the structure shown in Figure 33. Each line represents a specific value of the radial stress component.[14]

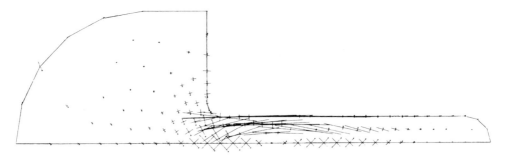

FIGURE 36.   Principal stresses in a section of the metal part of the structure (transverse force loading). Each cross represents the local magnitudes and orientations of the principal stresses.[14]

   The nodal point displacements and stresses calculated by the FEM program are presented in tabular printed form and in graphs of different kinds. These graphs are usually prepared by FEM postprocessors, computer programs that rearrange or adapt the output data according to the users' specifications. A few modes of graphical output are shown in the next figures. Figure 35 shows an iso stress line plot for a section of the metal part of the structure in Figure 33. The transverse (or radial) stress component (σ_r) is presented in lines of equal stress, where each line represents a certain stress value. Stress concentrations specifically are easily detected in these graphs, as one is evident in the head-to-stem connection region. Another method of giving an overall impression of the stresses is shown in Figure 36 for the same structure. The crosses represent the principal stress magnitudes and orientations in the plane of the section considered, calculated from the stress components (compare Figure 7), and give a good impression of the load transmittance through the structure. Note that near the stem-plug interface the maximal principal stress is in the axial direction, indicating that the bending stresses in the stem are much more significant than the transverse stresses at the interface. A last method of stress representation is presented in Figure 37, showing line graphs, stresses as occurring along a line in the structure. In this case, four stress components along a line in the bone close to the plug-bone interface are given: τ_{rz} is the shear stress at the interface in the longitudinal direction, σ_z is the axial direct stress (also shown as it would occur in the "natural" bone), σ_t is the circumferential direct stress (hoop stress), and σ_r is the transverse or radial stress.* Stress concentrations at the proximal side are apparent. A graphical method of presenting deformation, not shown here, is that of the

---

*   Because the structure is axisymmetric and the load in the case of this figure is too, the other stress components (τ_{rt} and τ_{tz}) are zero.

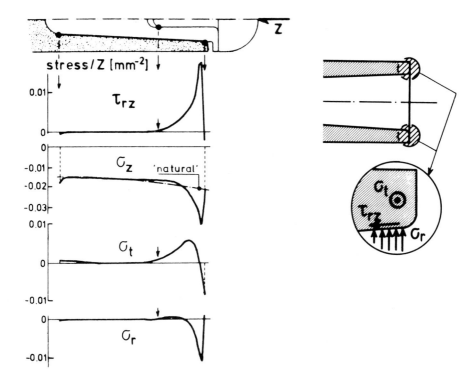

FIGURE 37.   Stresses as occurring along a line in the structure in bone at the plug-bone interface (axial loading case); σ_z is the axial direct stress; the meaning of the other components is illustrated in the inset. Stress concentrations in the proximal part are evident.[14]

deformed element mesh (represented by nodal point displacements) in comparison with the undeformed mesh; the deformations are usually exaggerated to make them visible with the naked eye.

Versatile features of FEM analysis when compared to experimental methods in cases that a structure is too complicated for closed-form theories are its potential for evaluating stresses throughout the structure, in and between all materials concerned, and for parametric analysis. Material properties, and loading and boundary conditions are readily varied to investigate their influences. An example is shown in Figure 38, where radial and hoop stresses in the plug material near the plug-bone interface are compared for three plug materials. Increasing stress values from Delrin® via UHMWPE to Silastic® are apparent. It is noteworthy that these increases are due to the increasing Poisson's ratio of the materials. The structure is loaded in the z direction. Both the radial and the hoop stresses in the plug depend on the compressibility of the plug material because the plug is restrained on all sides. Silastic® is almost incompressible and hence engenders high stresses in directions perpendicular to the load in this case.

Most FEM analysis published in the (hard-tissue) biomechanics literature as yet[12,13] assume linear elastic behavior. If materials are nonlinear elastic or viscoelastic and when geometrically nonlinear behavior* occurs, the application of the FEM is still well feasible, although more complicated. Nonlinear problems must be solved in step-by-step iteration procedures,[11]

---

*   In linear stress analysis, it is assumed that the deformations are small when compared to the significant dimensions of the structure. In that case, the deformations are at all times directly proportional to the magnitudes of the loads, and the principle of superposition holds (i.e., the stresses that result from a combined loading case are equal to the sums of the stresses resulting from each load separately). In the case that materials are nonlinear and/or if large deformations occur (geometrical nonlinearity), these principles become invalid.

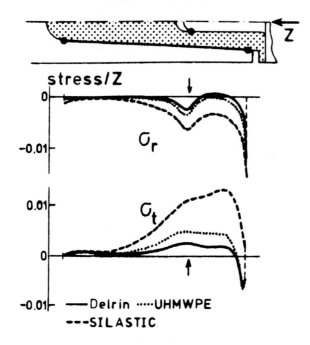

FIGURE 38.   A result of parametric analysis: radial and hoop stresses in the plug, close to the plug-bone interface for three different plug materials.[14]

increasing the loads from zero to their real values, which takes a large amount of computer time and space. Apart from its complexity, these kinds of analyses are often hampered by a lack of descriptive data on nonlinear material properties. As a consequence, materials are usually approximated as linear. Anisotropy and nonhomogeneity are easily taken into account in FEM models.

The most difficult steps in FEM analyses are the creation of the model and the interpretation of results in the light of the model limitations. Considerable expertise is needed to judge the accuracy, the validity, and the significance of the computer output. Accuracy is mainly a question of adequate (local) mesh density, as we have discussed previously. Convergence checks (repeating the calculations with a finer mesh) are always feasible, but sometimes expensive. Judging the validity of results requires a realistic assessment of model limitations. The FEM has the less desirable aspect of yielding little insight to the lay person. On the contrary, the amount and power of numerical computer data seem overwhelming to many, and the only "visible" feature is really a plot of the element mesh. One should remember that this drawing represents, of the four structural aspects, merely the geometry. Users of FEM results should also be keen on model assumptions with respect to the other aspects; an apparent visual agreement between model and reality does not guarantee validity in itself. Unfortunately, no rule of thumb can be given to assess whether model results can be regarded as realistic in an absolute sense, in a relative sense, or not at all. Often some results are more realistic than others, depending on how sensitive a certain stress value is to a certain assumption. For instance, it was shown[10] that when the interfaces in models of the femoral hip joint fixation structure are assumed rigidly bonded, and in reality they are loose, the most significant stem and bone stress results are still realistic, but the cement layer and interface stresses represent reality only in a rough relative sense. Sometimes it is intuitively obvious how a certain assumption affects a certain result (at least to the analyst). Sometimes this influence can be estimated from existing information or from simple analytical considerations. Often, however, this influence is simply unknown and should be established through appropriate research. This can be done by means of parametric analysis, for instance, comparing 2-D with 3-D models, isotropic assumptions with anisotropic assumptions, and so on. Another possibility is the use of verifying experiments on laboratory models, which may be simulated in FEM models to compare theoretical with experimental results.

Eventually, the stress results will be judged as to what they actually indicate. This is often a problem when the objective of an analysis entails the prevention of possible failure in connection with structural designs. It was, for instance, established in several analyses that cement failure in femoral hip joint prostheses is most likely to occur either on the proximal side or at the distal tip, where the cement stresses reach maximal values. It has also been established that proximal cement stresses increase on using a more flexible titanium alloy instead of a cobalt-chrome alloy for the stem, while the distal cement stresses decrease in that case. These facts can be established with structural analyses, however, the decision on what to use depends not only on the outcome of the analysis, but also on which failure mechanism one believes is the most destructive. Although there is more to this question of stem material than simply stated here, this is one example of how different analyses of the same structure can sometimes result in opposite recommendations. Another example concerns the interpretation of stresses in bone, where different authors may reach different conclusions based on comparable analyses, since so little is known about the biological reactions of bone to various kinds of stresses.

Possibilities of the FEM in biomechanics entail all potential uses of structural analysis, i.e., all cases where a quantitative assessment of stress is required, whether for the investigation of basic mechanisms or for the evaluation of designs and techniques. Its use is not limited to bone, but can be applied to soft tissues as well, although poor knowledge about collagenous tissue properties hamper its application as yet. As a numerical computer solution method, the FEM is also applied in biomechanics for problems of heat and mass transfer.[10,11]

Although its mathematical basis is highly complicated, the FEM is merely a tool of structural analysis, in which modeling is the true skill, requiring the most expertise. When this process is correctly understood, the FEM itself can be conveniently regarded as a "black box" by nonspecialists.

APPENDIX: PARAMETERS AND UNITS

The units in this chapter are based on the SI metric system, now used almost throughout the world. The base units are those of length (m, meter), mass (kg, kilogram), time (s, second), and of temperature (K, Kelvin). Derived units of interest in this chapter are those of force (N, Newton), pressure or stress (Pa, Pascal = N/m^2), and a supplementary unit is the radian (rad) for measuring angles (1 rad = $180/\pi$ degrees, $\pi \approx 3.14$). An excellent article on units and how they pertain to Biomechanics was published by Carter.[15]

Instead of the meter, the millimeter was used in this chapter. The unit of stress is then MPa (mega-Pascal = N/mm^2). The following parameters and abbreviations were used in the text:

Geometry

x,y,z	(mm)	Coordinates of a Cartesian coordinate system
r,ϕ,z	(mm,rad,mm)	Coordinates of a cylindrical coordinate system
α,β,ϕ	(rad)	Angles (sometimes expressed in degrees)
L,ℓ	(mm)	Lengths
$\Delta L,\Delta \ell$	(mm)	Changes in lengths
d	(mm)	Thickness
b,h	(mm)	Width and height of rectangular beams
e	(mm)	Distance from neutral axis to surface of a beam
R	(mm)	Radius

R_i, R_o	(mm)	Inner and outer radius of a hollow shaft
A	(mm^2)	Area
dA	(mm^2)	Infinitesimal area
I	(mm^4)	Second moment of inertia (SMI)
I_x, etc.	(mm^4)	SMI with respect to (around) x axis, etc.
I_{max}, I_{min}	(mm^4)	SMI with respect to principal inertia axes
I_p	(mm^4)	Polar SMI

Material properties

E	(MPa)	Young's modulus
G	(MPa)	Shear modulus
ν	—	Poisson's ratio

Loading

F, F_x, etc.	(N)	Force, force in x direction, etc.
M, M_x, etc.	(Nmm)	Moment, moment around x axis, etc.
M_t	(Nmm)	Torsion moment
M_{max}, M_{min}	(Nmm)	Bending moments around principal inertia axes
a,b,c	(mm)	Moment arms

Stresses

σ, σ_x, etc.	(MPa)	Stress, direct stress in x direction, etc.
τ, τ_{xy}, etc.	(MPa)	Shear stress, shear stress on x plane in y direction, etc.
σ_n	(MPa)	Stress normal to a particular plane
$\sigma_r, \sigma_t, \sigma_z$	(MPa)	Direct stresses in cylindrical coordinate system
$\tau_{rt}, \tau_{rz}, \tau_{tz}$	(MPa)	Shear stresses in cylindrical coordinate system
σ_b	(MPa)	Maximal bending stress in beams
τ_t	(MPa)	Maximal torsion stress in shafts
σ_1, σ_2	(MPa)	Principal stresses

Strains and deformations

ϵ, ϵ_x, etc.	(mm/mm)	Strain, strain in x direction, etc.
γ, γ_{xy}, etc.	(mm/mm)	Shear strain, shear strain in x plane in y direction, etc.
θ	(rad/mm)	Angle of twist per unit shaft length

Miscellaneous

EA	(Nmm^2/mm^2)	Axial rigidity of a bar
EI	(Nmm^4/mm^2)	Flexural rigidity of a beam
GI_p	(Nmm^4/mm^2)	Torsional rigidity of a shaft
AA,BB,CC	—	Section locations
FEM	—	Finite element method
$\int_A$	—	Integrated over area A

REFERENCES

1. **Chao, E. Y. S. and An, K. A.,** Perspectives in measurements and modelling of musculoskeletal joint dynamics, in *Biomechanics: Principles and Applications,* Huiskes, R., van Campen, D. H., and de Wijn, J. R., Eds., Martinus Nijhoff, The Hague, 1982, 1.
2. **Hayes, W. C. and Snyder, B.,** Toward a quantitative formulation of Wolff's law in trabecular bone, in *Mechanical Properties of Bone* (AMD - Vol. 45), Cowin, S. C., Ed., American Society of Mechanical Engineers, New York, 1981, 43.
3. **Huiskes, R., Janssen, J. D., and Slooff, T. J.,** A detailed comparison of experimental and theoretical stress-analyses of a human femur, in *Mechanical Properties of Bone* (AMD — Vol. 45), Cowin, S. C., Ed., American Society of Mechanical Engineers, New York, 1981, 211.

4. **Evans, F. G.,** *Mechanical Properties of Bone,* Charles C Thomas, Springfield, Ill., 1973.
5. **Reilly, D. T. and Burstein, A. H.,** The elastic and ultimate properties of compact bone tissue, *J. Biomech.,* 8, 393, 1975.
6. **Durelli, A. J.,** The difficult choice: evaluation of methods used to determine experimentally displacements, strains and stresses, *Appl. Mech. Rev.,* 30, 9, 1167, 1977.
7. **Lanyon, L. E.,** The measurement and biological significance of bone strain in vivo, in *Mechanical Properties of Bone* (AMD - Vol. 45), Cowin, S. C., Ed., American Society of Mechanical Engineers, New York, 1981, 93.
8. **Pauwels, F.,** *Biomechanics of the Locomotion Apparatus,* Springer-Verlag, Berlin, 1980.
9. **Crippen, T. E. and Huiskes, R.,** unpublished data, 1980.
10. **Huiskes, R.,** Some fundamental aspects of human joint replacement; analyses of stresses and heat-conduction in bone-prosthesis structures, *Acta Orthop. Scand. Suppl.,* 185, 1979.
11. **Zienkiewicz, O. C.,** *The Finite Element Method,* 3rd ed., McGraw-Hill, London, 1977.
12. **Huiskes, R. and Chao, E. Y.,** A survey of finite element methods in orthopaedic biomechanics: the first decade, *J. Biomech.,* 16, 385, 1983.
13. **Gallagher, R. H., Simon, B. R., Johnson, P. C., and Gross, J. F., Eds.,** *Finite Elements in Biomechanics,* John Wiley & Sons, New York, 1982.
14. **Huiskes, R., van Heck, J., Walker, P. S., Green, D. J., and Nunamaker, D.,** A three-dimensional stress analysis of a new finger-joint prosthesis fixation system, in *Proc. Finite Elements Biomechanics,* Vol. 2, Simon, B. R., Ed., University of Arizona, Tucson, 1980, 749.
15. **Carter, D. R.,** SI: the international system of units, in *Basic Biomechanics of the Skeletal System,* Frankel, V. H. and Nordin, M., Eds., Lea & Febiger, Philadelphia, 1980, 1.
16. **Cowin, S. C., Ed.,** *Mechanical Properties of Bone,* AMD - Vol. 45, American Society of Mechanical Engineers, New York, 1981.
17. **Feinberg, B. N. and Fleming, D. G., Eds.,** *CRC Handbook of Engineering in Biology, Section B,* Instruments and Measurements, Vol. 1, CRC Press, Boca Raton, Fla., 1978.
18. **Frankel, V. H. and Burstein, A. H.,** *Orthopaedic Biomechanics,* Lea & Febiger, Philadelphia, 1970.
19. **Frankel, V. H. and Nordin, M.,** *Basic Biomechanics of the Skeletal System,* Lea & Febiger, Philadelphia, 1980.
20. **Fung, Y. C.,** *Biomechanics, Mechanical Properties of Living Tissues,* Springer-Verlag, New York, 1981.
21. **Lai, W. M., Rubin, D., and Krempl, E.,** *Introduction to Continuum Mechanics,* Pergamon Press, Oxford, 1978.
22. **Radin, E. L., Simon, S. R., Rose, R. M., and Paul, J. L.,** *Practical Biomechanics for the Orthopaedic Surgeon,* John Wiley & Sons, New York, 1979.
23. **Simmons, D. J. and Kunin, A. S., Eds.,** *Skeletal Research,* Academic Press, New York, 1979.
24. **Timoshenko, S. P. and Goudier, J. N.,** *Theory of Elasticity,* McGraw-Hill, Kogahucha, 1970.
25. **Wainwright, S. A., Biggs, W. D., Currey, J. D., and Gosline, J. M.,** *Mechanical Design in Organisms,* Edward Arnold, London, 1976.
26. **White, A. A. and Panjabi, M. M.,** *Clinical Biomechanics of the Spine,* Lippincott, Philadelphia, 1978.
27. **Yamada, H.,** *Strength of Biological Materials,* Robert E. Krieger, Huntington, N.Y., 1973.

Chapter 5

DYNAMIC ANALYSIS OF HUMAN BONES

G. Van der Perre

TABLE OF CONTENTS

I.   Introduction 100

II.   Elements of Dynamic Analysis 100
      A.   Dynamic vs. Static Analysis 100
      B.   Relation Between Free Vibration Motion and Dynamic Response 102
            1.   Undamped Free Vibration 102
            2.   Single Degree of Freedom (SDOF) Systems 103
            3.   Impulse Response 104
            4.   Frequency Response 104
            5.   Relation Between Frequency Response, Dynamic Response,
                  and Impulse Response 107
            6.   Relation Between Frequency Response and Free Vibration
                  Parameters: Undamped Case 110
            7.   Relation Between Frequency Response and Free Vibration
                  Parameters: Damped Case 111
            8.   Discussion of the SDOF Frequency Response Function 112
            9.   Laplace Transformation and Transfer Function 114
      C.   Dynamic Analysis of Multiple Degrees of Freedom (MDOF) Systems .. 115
            1.   Degrees of Freedom 115
            2.   Undamped Free Vibration of a MDOF System: Modal
                  Parameters 116
            3.   Relation Between Free Vibration Modal Parameters and
                  Dynamic Response of MDOF Systems 118
                  a.   Point Impulse Response and Frequency Characteristics
                        of MDOF Systems 120
                  b.   Influence of Damping — Proportional and Nonpro-
                        portional Damping 122
                  c.   Response to Actual Dynamic Loading: Modal
                        Coordinates 124
            4.   Discretization and Modeling 126
      D.   Frequency Response Testing 126

III. In Vivo Testing of Long Bones 132
      A.   Steady-State Vibration Tests In Vivo — Beam Models 133
            1.   Resonant Frequency Determinations 133
            2.   Driving Point Impedance Tests 134
      B.   Generalized MDOF Dynamic Analysis of Long Bones 136
            1.   Finite Element Modeling 137
            2.   Modal Analysis 141
                  a.   Dynamic Characteristics of the Free Tibia 142
                  b.   Modal Analysis of Human Tibias In Vivo 146
      C.   Monitoring of Fracture Healing by Dynamic Response Analysis 149
      D.   Diagnosis of Osteoporosis 154

IV.   Analysis of the Dynamic Response in Actual Loading Conditions 155
      A.   Dynamic Analysis of the Femur 155
      B.   Dynamic Analysis of the Skull 155
      C.   Dynamic Analysis of the Human Spine 155

V.   Dynamic Analysis of Bone Implant Assemblies 156
     A.   Fracture Fixation Plates on Tibias 156
     B.   Total Hip Prostheses in Femurs 156

References 158

I. INTRODUCTION

Within the area of mechanics, dynamic analysis is the discipline dealing with the displacements (motions) by which structures respond to dynamic, i.e., time varying, loading. Dynamic analysis of human bones is interesting from two application points of view: (1) In vivo testing of whole bone mechanical properties by dynamic testing methods and (2) predicting of the response of bones to real life dynamic loading conditions. The basic concept uniting both applications is the impulse response concept: a structure responds to an external impulse by its own free vibration motion, which is determined by the spatial distribution of mass, stiffness, and damping properties over the structure and by the boundary conditions, i.e., the connections with the surroundings.

In the first application, frequency response testing is made in order to determine the free vibration characteristics of the structure and hence the physical properties of the structure and/or the boundary conditions. In bone biomechanics, this application includes clinical assessment of bone (e.g., fracture healing, osteoporosis) or joint properties.

In the second application, free vibration characteristics (obtained by model calculations or from frequency response tests) are used in the computation of the response of bones in specific dynamic loading situations. Typical examples are impact and vibration, but even in normal walking, dynamic effects could play. A very specific application is the analysis of bone conducted vibrations in sensory feedback and in ear stimulation.

In Section II of this chapter, the basic elements of theoretical and experimental dynamic analysis are explained. Applications of dynamic analysis in biomechanics are rather recent, and more particularly in connection with the real life dynamic response analysis application, we are in a stage where we can discuss potentialities rather than report on finished studies. Therefore, the first section is rather extensive. The theoretical development is partly based upon the book by Clough and Penzien.[1] For measurement and instrumentation details, we refer to the book edited by Bruël and Kjaer.[2]

Section III gives a comprehensive review of the research made on in vivo testing of long bones, mainly in connection with fracture healing and osteoporosis. A number of teams have been working on this subject for a decade. Clinical application can be expected in the near future. The two last sections deal briefly with studies in connection with real life dynamic response (spine, skull, and femur, Section IV) and with bone-implant assemblies (fracture fixation plates and hip prostheses, Section V).

II. ELEMENTS OF DYNAMIC ANALYSIS

A. Dynamic vs. Static Analysis (Figure 1)

The static equilibrium displacement, x, of a spring-connected mass (see Figure 1A) under

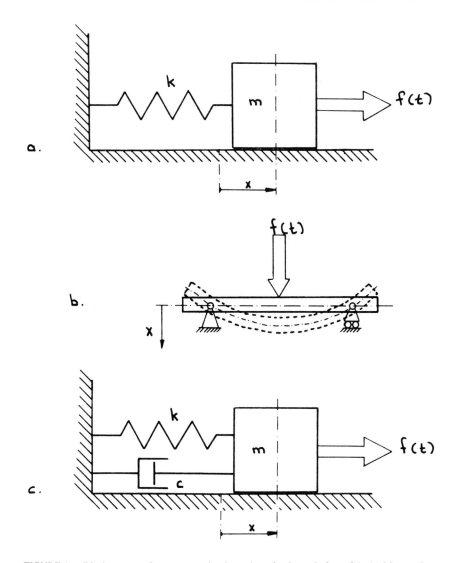

FIGURE 1. Displacement of a system under the action of a dynamic force f(t). (a) Mass-spring system; k, spring constant; m, mass; x, displacement. (b) beam in bending deformation; x, deflection. (c) mass-spring-dashpot system; c, damping constant.

the action of a constant force, f_o, obeys the simple formula

$$kx = f_o \tag{1}$$

in which k is the spring constant (spring stiffness). Similarly, a constant force, f_o, acting upon the beam of Figure 1B results in a static deflection, x, obeying the same formula (see Equation 1) in which k indicates the transverse beam stiffness.

On the other hand, if a time varying force, f(t), is applied, the motion of the mass, m, obeys the formula

$$m\ddot{x} + kx = f(t) \tag{2}$$

or

$$f_I + f_S + f(t) = 0 \tag{3}$$

which expresses the dynamic equilibrium between the external force, f(t), the spring force, $f_s = -kx$, and the "inertial force", $f_I = -m\ddot{x}$. Equation 2 is the differential equation of motion of the system. The deflection, x, of the beam (see Figure 1B) follows a similar equation, expressing the equilibrium between external, elastic, and inertial forces. Dynamic analysis differs from static analysis, not only in that time varying forces are dealt with, but also in that mass-inertial forces are taken into account.

If the parameters of the systems (see Figure 1A) and (see Figure 1B) are such that the amount of moving mass and the accelerations are small in comparison with external forces and stiffness, inertial forces can be neglected and Equation 2 reduces to

$$kx(t) = f(t) \tag{4}$$

This is the "quasistatic" approach, which considers motion as a sequence of static equilibrium states. From the energy point of view, this means that the work done by the force is assumed to be directly transformed into elastic energy; kinetic energy is neglected. The quasistatic approach deals mainly with peak strains, x(t), associated with peak forces, f(t), and is currently applied in experimental and mathematical strain analysis of bones and implants, e.g., in normal gait loading conditions.

A third characteristic of dynamic analysis is that it includes damping phenomena. In Figure 1C viscous damping is symbolized by a dashpot, which resists motion by a force proportional to the velocity, $\dot{x}$, with proportionality coefficient, c, the damping constant. The dynamic equilibrium equation for this system is

$$m\ddot{x} + c\dot{x} + kx = f(t) \tag{5}$$

or

$$f_I + f_D + f_s + f(t) = 0 \tag{6}$$

$f_D = -c\dot{x}$ is the damping resistance.

Again, a system can be approached as visco-elastic, with neglection of mass inertial effects. This approach is common in impact tests on human bones, in which experiments are run in conditions of roughly constant $\dot{x}$.[3] With this approximation, the equilibrium equation becomes

$$c\dot{x} + kx = f(t) \tag{7}$$

Since, if $\dot{x}$ remains constant, no kinetic energy is produced, work by the force, f(t), is transformed directly into elastic energy and heat.

The general dynamic response of a system, as the mass-spring-dashpot system (see Figure 1C, Equations 5 and 6), is intimately related to the system's free vibration characteristics.

B. Relation Between Free Vibration Motion and Dynamic Response
1. Undamped Free Vibration
The free vibration (i.e., the movement without any external exciting force) of a mass-spring system (see Figure 2A) is governed by the condition of equilibrium between elastic and inertial force:

$$f_s + f_I = 0 \quad \text{or} \quad m\ddot{x} + kx = 0 \tag{8}$$

The harmonic motion, $x = X \sin(\omega_n t + \varphi)$ (see Figure 2C), obeys this condition if ω_n

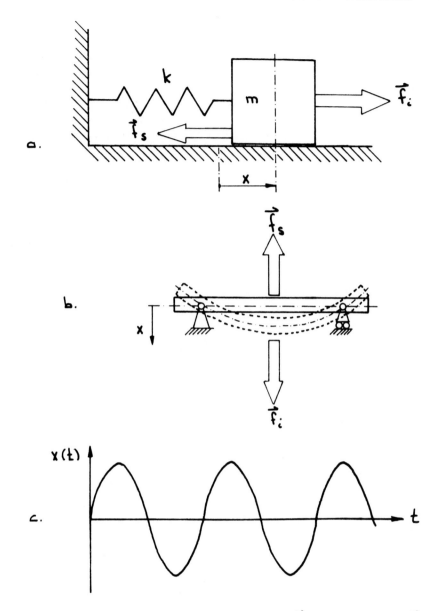

FIGURE 2.   Free vibration motion of an undamped system; $\vec{f}_s$, elastic (stiffness) force; $\vec{f}_i$, inertial force. (a) Mass-spring system; (b) beam in bending deformation; (c) harmonic motion $x(t) = X \sin(\omega_n t + \varphi)$ ($\varphi = 0$).

$= \sqrt{k/m}$; ω_n is the "natural frequency". X and φ depend upon initial conditions. This reasoning can again be extended to the beam of Figure 2B: if its deflection shape is known and unique, its vibration can be characterized, e.g., by the movement of the middle point, $x_M = X_M \sin(\omega_n t + \varphi)$, in which $\omega_n = \sqrt{K/M}$. K is the "generalized stiffness", representing the elastic resistance of the structure to a deformation with this particular shape, and M is the "generalized mass", representing the structure's inertial resistance to accelerations inherent to this movement. Hence, for any free vibration, ω_n^2 is proportional to "stiffness" and inversely proportional to "mass".

2. Single Degree of Freedom (SDOF) Systems

Systems, the motion of which can be described by one single parameter, as is obvious for the rectilinear motion of a mass particle and as we have assumed for a deflecting beam (see Figures 1 and 2), are called "single degree of freedom" (SDOF) systems. The discussion in Section II.B is confined to SDOF systems. The problem of multiple degrees of freedom connected with continuous structures ("distributed parameter systems") is dealt with in Section II.C.

3. Impulse Response

The physical relation between free vibration and dynamic response is based upon the impulse response concept. If the mass of Figure 2A or beam of Figure 2B (still assuming it only has one single mode of vibration) is excited by an extremely short force impulse, it responds by its free vibration movement. The free vibration movement excited by a unit impulse, $\delta(t)$, the latter being defined by lim

$$\epsilon \rightarrow 0 \int_{-\epsilon}^{\epsilon} \delta(t)dt = 1$$

is called the impulse response function, $h(t)$.

An arbitrary dynamic force, $f(t)$, can be considered as a sequence of impulses with infinitesimal with $d\tau$ and height $f(\tau)$ (Figure 3). For the calculation of the result, a displacement, $x(t)$, the superposition principle (valid for linear differential equations), is applied which states that the effect of simultaneously superimposed actions is equal to the sum of the effects of each individual action. Mathematically, this principle can be written (see Figure 3)

$$x(t) = \int_{-\infty}^{t} f(\tau)\, h(t - \tau)d\tau \qquad (9)$$

This mathematical operation, by which $x(t)$ is determined from $f(t)$ and the impulse response function $h(t)$ is called convolution, is represented symbolically by

$$x(t) = f(t) \star h(t) \qquad (10)$$

4. Frequency Response

The impulse response superposition principle, although physically clear, leads to the mathematically complicated convolution operation.[7] However, in the particular case where $f(t)$ is a harmonic forcing function, $f(t) = F_\omega \sin \omega t$ (ω and F_ω being, respectively, the angular frequency and amplitude of this forcing function), Equation 2, $mx + kx = f(t)$, can be solved directly, without using the impulse response concept. The solution of Equation 2 consists of the general solution x_T to the homogeneous Equation 8 (i.e., the free vibration movement) plus a particular solution x_S to Equation 2 as such.

$$x = x_T + x_S = X_T \sin (\omega_n t + \varphi_n) + X_S \sin (\omega t + \psi_\omega) \qquad (11)$$

When damping is present, the free vibration term becomes (see Section II.B.7) $x_T = \exp^{-\sigma t} X \sin(\omega t + \varphi)$ (see Equation 11), i.e., a "transient" term, which vanishes with time. Hence, after a certain period of time, only the particular solution remains, which is called the "steady-state response" term,

$$x_S = X_S \sin (\omega t + \psi_\omega) \qquad (12)$$

In view of further generalization, this solution is represented in its complex notation:

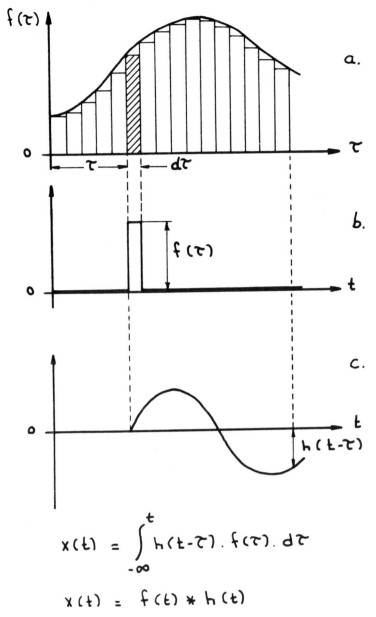

$$x(t) = \int_{-\infty}^{t} h(t-\tau) \cdot f(\tau) \cdot d\tau$$

$$x(t) = f(t) * h(t)$$

FIGURE 3.   Superposition principle. (3a and 3b) Force f(τ) as a sequence of impulses f(τ) dτ; (3c) impulse response h (t − τ). (3c) Superposition principle (convolution):

$x(t) = t \int_{-\infty}^{t} h(t - \tau) \times f(\tau)\, d\tau.$

$$x_s = 1/2(\underline{X}_s\, e^{j\omega t} + \underline{X}_s{}^{\star}\, e^{-j\omega t}) \tag{13}$$

$$(\underline{X}_s = \frac{X_s}{j}\, e^{j\psi\omega}, \underline{X}_s{}^{\star} = \text{complex conjugate of } \underline{X}_s)$$

In Equation 13, the movement, x(t), is represented by two vectors of length, $X_{S/2}$, rotating in opposite directions with angular velocity, ω. The imaginary parts of the vector pair always cancel each other (Figure 4). This solution, substituted into Equation 2, yields (with the same complex notation used for the forcing function):

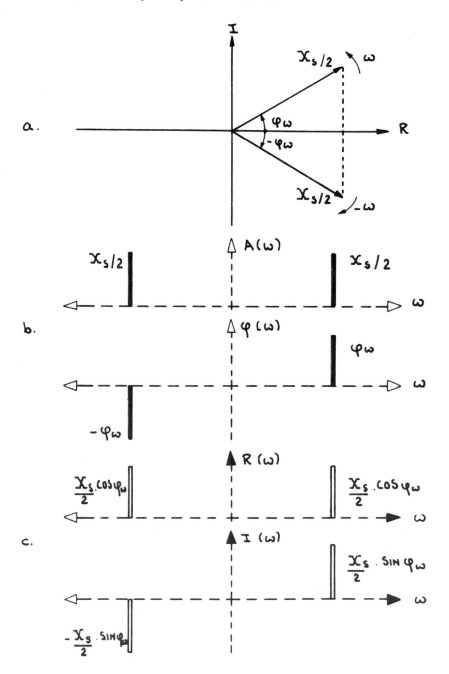

FIGURE 4.   Complex representation of steady state vibration. (a) Rotating vector representation; I, imaginary axis; R, real axis; ω, circular frequency; X_s, amplitude; φ_ω, phase angle. (b) Amplitude and phase representation; $A(\omega)$, amplitude; $\varphi(\omega)$, phase angle. (c) Real and imaginary part representation, $R(\omega)$, real part; $I(\omega)$, imaginary part.

$$\underline{X}_S \, (-m\omega^2 + k) = \underline{F}$$

and hence

$$\frac{\underline{X}_S(\omega)}{\underline{F}(\omega)} = \frac{\underline{X}_S{}^\star(\omega)}{\underline{F}{}^\star(\omega)} = \frac{1}{-m\omega^2 + k} = \underline{H}(\omega) \qquad (14)$$

$H(\omega)$ is the frequency response function. $\underline{H}(\omega)$ is real in the undamped case. In the damped case, the frequency response function consists of two conjugate parts (ω is a positive quantity in the expressions):

$$\frac{\underline{X}_S(\omega)}{\underline{F}(\omega)} = \frac{1}{-m\omega^2 + j\omega C + k} = \underline{H}(\omega) \tag{15}$$

and

$$\frac{\underline{X}_S^\star(\omega)}{\underline{F}^\star(\omega)} = \frac{1}{-m\omega^2 - j\omega C + k} = \underline{H}^\star(\omega) \tag{16}$$

If ω is allowed to take negative values, Equation 15 for $\omega = |\omega|$ and $\omega = -|\omega|$ expresses both parts.

Since either of the conjugate parts of $\underline{F}(\omega)$, $\underline{X}_S(\omega)$, and $\underline{H}(\omega)$ contain the amplitude as well as the phase information, it is sufficient in the discussion to look only at the part with positive ω, thus to represent f(t) and x(t) as

$$f(t) = \underline{F}(\omega) \, e^{j\omega t} \qquad\qquad \underline{F}(\omega) = |F(\omega)| \, e^{j\theta_\omega}$$
$$x_S(t) = \underline{X}_S(\omega) \, e^{j\omega t} \qquad\qquad \underline{X}_S(\omega) = |X_S(\omega)| \, e^{j\psi_\omega}$$
$$\underline{H}(\omega) = \frac{\underline{X}_S(\omega)}{\underline{F}(\omega)} = \frac{|X_S(\omega)|}{|F(\omega)|} \, e^{j(\psi_\omega - \theta_\omega)} = |H(\omega)| \, e^{j\varphi_\omega}$$

The part for $-j\omega$ can simply be seen as the value of each function for the associated negative value of ω: same amplitude, opposite phase angle.

In the undamped case (see Equation 14),

$$|H(\omega)| = \frac{1}{-m\omega^2 + k} = \frac{1/m}{-\omega^2 + k \, m} \tag{17}$$
$$\varphi_\omega = 0$$

In the damped case (see Equations 15 and 16),

$$\left. \begin{array}{l} |H(\omega)| = \dfrac{1/m}{\sqrt{(\omega_n^2 - \omega^2)^2 + C\omega^2/m^2}} \\[2em] \varphi_\omega = -\arctan \dfrac{\omega C}{m(\omega_n^2 - \omega^2)} \end{array} \right\} \tag{18}$$

In Figure 6, $|H(\omega)| \, xk = X_S(\omega)/X_{stat}$, the ratio of the steady-state amplitude to the static displacement, and φ_ω, the phase angle, are plotted vs. ω/ω_n, for the undamped case and a number of damped cases. These graphs are discussed further (see Section II.B.8), but we will first extend their physical meaning to the general dynamic response problem.

5. Relation Between Frequency Response, Dynamic Response, and Impulse Response

The complex frequency response function $\underline{H}(\omega)$ represents the system's steady-state response to a harmonic forcing function with frequency ω. Which is the relation with the dynamic response to an arbitrary force function, f(t)? A function, f(t), can be represented by

$$f(t) = 1/2\pi \int_{-\infty}^{\infty} F(\omega)e^{j\omega t} \, d\omega \qquad (19)$$

$F(\omega)$ is determined from $f(t)$ by the "complex Fourier transformation":

$$F(\omega) = \int_{-\infty}^{\infty} f(t) \, e^{-j\omega t} \, dt \qquad (20)$$

The complex Fourier transformation of $f(t)$ represents the function as a summation of an infinite number of harmonics, $F(\omega) \, d\omega$, which are represented by complex conjugates in just the same way as in Equation 13. It is again sufficient to look at the part with positive ω: $F(\omega) = |F(\omega)| \, e^{j\varphi\omega}$. By the Fourier transformation, the time domain representation $f(t)$ is transformed into the frequency domain representation, $F(\omega)$ (Figure 5).

Both forces $f(t)$ and displacement $x(t)$ can be Fourier transformed. Since the differential Equations 5 and 6 must hold for each harmonic contained in $x(t)$ and $f(t)$, Fourier transformation of Equations 5 and 6 results in

$$(-\omega^2 m + j\omega C + k) \, \underline{X}(\omega) = \underline{F}(\omega)$$

which is identical to the steady-state response Equation 15. The Fourier transformation transforms the differential equation into an algebraic equation.

The complex frequency response function,

$$\underline{H}(\omega) = \frac{1/m}{-\omega^2 + j\omega C/m + k/m} = \frac{\underline{X}(\omega)}{\underline{F}(\omega)} \qquad (21)$$

which was already defined and discussed in Equations 15 to 18, is now defined in a general way: it is the ratio of the Fourier transforms of $x(t)$ and $f(t)$.

The problem of finding $x(t)$ for an arbitrary $f(t)$ (for this SDOF) can now in principle be solved by determining $\underline{H}(\omega)$, Fourier transforming $f(t)$ to $\underline{F}(\omega)$ and applying the relation:

$$\underline{X}(\omega) = \underline{H}(\omega) \times \underline{F}(\omega) \qquad (22)$$

Then $x(t)$ can be determined by the reverse transformation

$$x(t) = 1/2\pi \int_{-\infty}^{\infty} \underline{X}(\omega) \, e^{j\omega t} \, d\omega \qquad (23)$$

This way $x(t)$ is represented by a combination of steady state responses to harmonic forces. When comparing Equation 22 with Equation 8, we see that the convolution in the time domain

$$x(t) = f(t) \star h(t)$$

of the force function, $f(t)$, with the (free vibration) impulse response function, $h(t)$, is replaced by a simple multiplication in the frequency domain of the force function, $\underline{F}(\omega)$, with the frequency response function, $\mathcal{H}(\omega)$:

$$\underline{X}(\omega) = \underline{F}(\omega) \times \underline{H}(\omega)$$

The "Fourier convolution theorem" states that the Fourier transform of a convolution is a

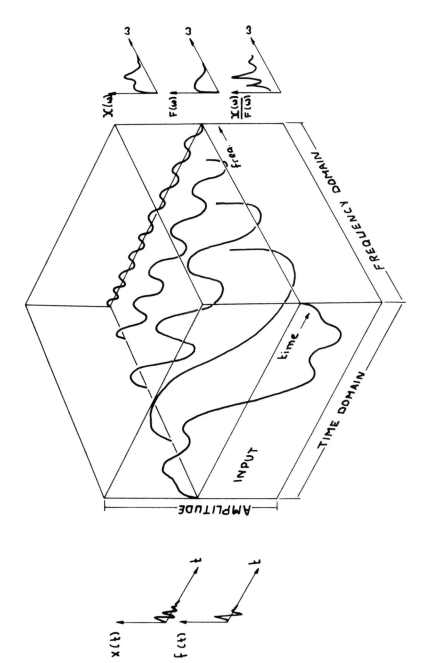

FIGURE 5. Time domain-frequency domain relation. x(t), f(t), time domain representation of displacement and force; X(ω), F(ω), frequency domain representation of displacement and force amplitude; H(ω) = X(ω)/F(ω), amplitude of frequency response. (From Péters, J., *Proc. Int. Seminar Modal Analysis*, Van Brussel, H. and Snoeys, R., Eds., Katholieke Universiteit Leuven, 1977, 18. With permission.)

multiplication.* This theorem, applied to Equation 8, means that the frequency response function $\mathbf{H}(\omega)$ is the Fourier transform of the impulse response function h(t).

Indeed, since the impulse response function h(t) is the response to a unit impulse $\delta(t)$, applying Equation 22, the Fourier transform of the impulse response function $\underline{\mathcal{H}}(\omega)$ is equal to

$$\underline{\mathcal{H}}(\omega) = \underline{\Delta}(\omega) \times \underline{\mathbf{H}}(\omega) \tag{24}$$

Since $\Delta(\omega)$, the Fourier transform of the unit impulse, is constant and equal to 1,

$$\underline{\mathcal{H}}(\omega) \equiv \underline{\mathbf{H}}(\omega) \tag{25}$$

6. Relation Between Frequency Response and Free Vibration Parameters: Undamped Case

In the undamped case, the impulse response is a free vibration motion (starting at $t = 0$, the instant the impulse is given):

$$x(t) = X \sin (\omega_n t + \varphi) \; (t > 0)$$

X and φ can be found from initial conditions: at $t = 0$, $x = 0$, hence $\varphi = 0$; $\dot{x}(t = 0)$ obeys to the impulse momentum law

$$m\dot{x}_{(t=0)} = 1, \quad \text{thus} \quad \dot{x}_{(t=0)} = 1/m, \text{ from which } X = \frac{1}{m\omega_n}$$

Hence, according to Equation 13, one would expect the complex Fourier representation of the undamped impulse response $\mathbf{H}(\omega)$ to consist of the discrete values

$$\underline{\mathbf{H}}(\omega_n) = \frac{1}{j2m\,\omega_n} \quad \text{and} \quad \underline{\mathbf{H}}(-\omega_n) = \frac{1}{-j2m\,\omega_n} \tag{26}$$

and to be zero elsewhere.

This, however, would correspond to perpetual harmonic motion (for $-\infty < t < \infty$), whereas the impulse response itself is the harmonic motion multiplied by a "window function" which is equal to zero below $t = 0$ and equal to 1 from $t = 0$ to ∞. Following the Fourier convolution theorem, the frequency response is then the perpetual motion transform (the discrete values, Equation 26), convoluted by the window transform (which is the hyperbolic function, $1/j\omega$). This way, the undamped frequency response function of Figure 6 is obtained:

$$\underline{\mathbf{H}}(\omega) = \frac{\dfrac{1}{j\,2m\,\omega_n}}{j(\omega - \omega_n)} + \frac{\dfrac{-1}{j2m\,\omega_n}}{j(\omega + \omega_n)} \tag{27}$$

which can also be obtained by partial fraction expansion of Equation 14. In abbreviated notation,

$$\underline{\mathbf{H}}(\omega) = \frac{\underline{\mathbf{A}}}{j(\omega - \omega_n)} + \frac{\underline{\mathbf{A}}^\star}{j(\omega + \omega_n)} \tag{28}$$

$\mathbf{A}$ and $\mathbf{A}^\star$ are the "residue values" of the frequency response function. Physically they represent the amplitude and phase of the free vibration excited by a unit impulse.

---

* And vice versa.

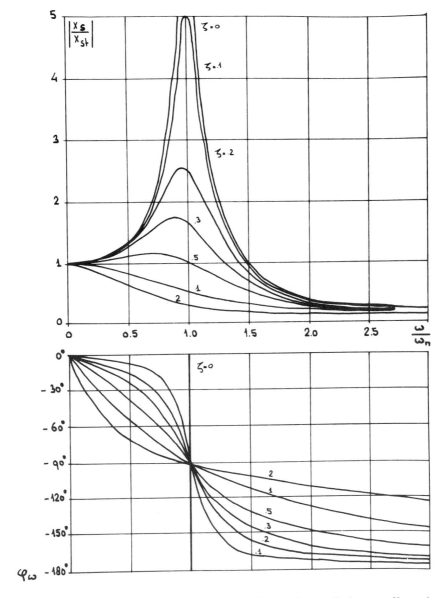

FIGURE 6. Frequency response of a SDOF system. X_S, steady state displacement; X_{St}, static displacement; $|X_S/X_{Stat}|$, dynamic magnification factor; φ_ω, phase angle; ω, circular frequency of forcing function; ω_n, natural frequency; ζ, damping coefficient.

7. Relation Between Frequency Response and Free Vibration Parameters: Damped Case

The solution to the damped free vibration equation,

$$m\ddot{x} + C\dot{x} + kx = 0$$

has the form

$$x = C_1 e^{s_1 t} + C_2 e^{s_2 t} \tag{29}$$

in which

$$s_{1,2} = -\frac{C}{2m} \pm \sqrt{(C/2m)^2 - k/m} \qquad (30)$$

If $C = 2\sqrt{km} = C_C$, the "critical damping value", there is only one solution, and $x(t)$ is an exponentially decaying function:

$$x(t) = X_o \, e^{-C/2mt} \qquad (31)$$

For C values below C_C, s_1 and s_2 are complex conjugates, and the solution can be written:

$$x = X \, e^{-\sigma t} \sin(\omega_d + \varphi) \qquad (32)$$

where $\sigma = C/2m = \zeta\omega_n$, in which ζ is the damping ratio, defined by $\zeta = C/C_m$, $\omega_d = \omega_n \sqrt{1 - \zeta^2}$, the "damped natural frequency".

If $C > C_C$, the solution is a sum of two exponentials.

We further discuss the "underdamped" case $C < C_C$. The free vibration is now an harmonic oscillation with frequency ω_d, the amplitude of which is exponentially decaying with time, mechanical energy being dissipated into heat (see Figure 7).

In the presence of damping, the impulse response is a damped free vibration movement starting at $t = 0$. This can be seen as an harmonic oscillation with ω_d as frequency, multiplied by an "exponentially decaying" window. Hence, the frequency response function will now consist of the "residue values" convoluted by the Fourier transform of this exponentially decaying window, which is $1/\sigma + j\omega$

$$H(\omega) = \frac{\mathbf{A}}{j(\omega - \omega_d) + \sigma} + \frac{\mathbf{A}^\star}{j(\omega + \omega_d) + \sigma} \qquad (33)$$

where $\underline{\mathbf{A}} = 1/2mj\omega_d$, since the oscillation has now frequency ω_d instead of ω_n.

8. Discussion of the SDOF Frequency Response Function (see Figure 6)

If the frequency of the forcing function $\omega = 0$, then $X = F/k$, the static displacement X_{ST}. We introduce the ratio $X(\omega)/X_{ST}$, which is called the "dynamic magnification factor" (equal to $k \times |H(\omega)|$) as it represents the amplification of the displacement response X by dynamic effects. This factor, together with the phase angle φ, is plotted in Figure 6 vs. ω/ω_n for various ζ values from $\zeta = 0$ to $\zeta = 2$ (overdamped).

The graphs illustrate the way in which the ratios ω/ω_n and $\zeta = C/2m\omega_n$ (thus, ω_n, the undamped natural frequency, being the relevant parameter) influence the frequency response, i.e., the steady-state response to an harmonic force as well as the Fourier transform of the impulse response function. The former meaning refers to experimental steady-state vibration tests, while the latter refers to the response to a general dynamic force, since

$$\underline{X}(\omega) = \underline{H}(\omega) \times \underline{F}(\omega) \qquad (34)$$

and

$$x(t) = 1/2\pi \int_{-\infty}^{\infty} \underline{X}(\omega) \, e^{-j\omega t} d\omega \qquad (35)$$

In all graphs, the left-hand part ($\omega/\omega_n \ll 1$) represents the "quasistatic" (almost frequency independent) response, $X(\omega) \approx X_{ST} = F/k$, only dependent upon the spring stiffness. The right-hand part ($\omega/\omega_n \gg 1$) represents the quasiinertial response, $X/F \approx -1/m\omega^2$, which is mass determined and decreases with ω^2.

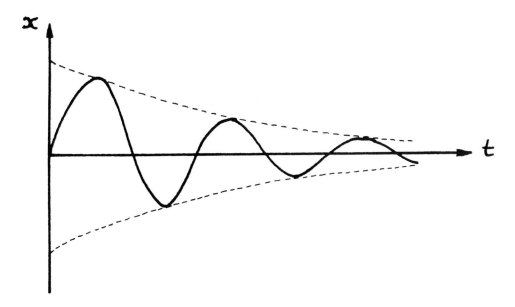

FIGURE 7.   Subcritically damped free vibration motion $x(t) = X e^{-\sigma t} \sin(\omega_{dt} + \varphi)$ $(\varphi = 0)$.

In the undamped case, X/X_{ST} becomes infinite at $\omega = \omega_n$ (resonance) and then it changes sign: left from resonance x is in phase with f, right from resonance x is in antiphase with f, or φ is 180°.

The (subcritically) damped graphs, when compared with the undamped case, reveal the following differences:

1.   The "peak value" $\underline{\omega}_p$ at which X/X_{ST} reaches its maximum is

$$\omega_p = \sqrt{1 - 2\zeta^2} \omega_n \tag{36}$$

2.   Increasing ζ lowers the X/X_{ST} graphs over the whole frequency range.
3.   The phase lag φ increases gradually with ω/ω_n and reaches 90° at $\omega/\omega_n = 1$. At $\omega = \omega_n$, inertial force and elastic force equilibrate each other, hence there should be equilibrium between external force and damping force as well. Since the damping force is in phase with velocity and displacement lags behind 90° on velocity, displacement lags behind by 90° on external force. Beyond ω_n, φ further increases to 180°. At $\varphi = 180°$, external force is in equilibrium with inertial force.

Finally, the critically damped and overdamped graphs show a steadily decreasing X/X_S, as an effect of the viscous resistance, increasing with velocity. The phase lag tends to 90° over the whole frequency range: external force in equilibrium with damping force.

Remarks — The frequency response function $\underline{H}(\omega)$ discussed above is the displacement frequency response function (also called dynamic compliance). Since in most experiments acceleration is measured instead of displacement, the acceleration frequency response function is also of interest (also called inertance). The latter is obtained by multiplying $\underline{H}(\omega)$ by $(-\omega^2)$. As a result of this multiplication, the peak value ω/ω_n at which the acceleration amplitude reaches its maximum is shifted to a higher value:

$$\omega_p = \frac{\omega_n}{\sqrt{1 - \zeta^2}} \tag{37}$$

In some practical vibration studies, mechanical impedance, $\underline{Z}(\omega) = \underline{F}(\omega)/\underline{V}(\omega)$, is monitored,

$\underline{V}(\omega)$ being the velocity Fourier transform. Mechanical impedance amplitude is the force amplitude needed to excite a velocity response of unit amplitude, hence expressing the system's resistance to movement.

$$\underline{Z}(\omega) \ = \ \frac{1}{j\omega\underline{H}(\omega)}$$

Other definitions of frequency response functions are

Dynamic stiffness: $\dfrac{\underline{F}(\omega)}{\underline{X}(\omega)}$

Dynamic mass: $\dfrac{\underline{F}(\omega)}{\underline{A}(\omega)}$ ($\underline{A}(\omega)$ is the Fourier transform of the acceleration)

Mobility: $\dfrac{\underline{V}(\omega)}{\underline{F}(\omega)}$

9. Laplace Transformation and Transfer Function

For the mathematical analysis of damped systems, the Laplace transform is appropriate, defined as

$$\underline{F}(s) \ = \ \int_0^\infty f(t) \ e^{-st} \ dt \tag{38}$$

with $s \ = \ \sigma \ + \ j\omega$.

The inverse transformation to the time domain is given by

$$f(t) \ = \ \frac{1}{2\pi j} \int_{\sigma-j\infty}^{\sigma+j\infty} \underline{F}(s) \ e^{st} \ ds \tag{39}$$

If σ is negative, the inverse transformation represents the function as a combination of damped harmonics with exponential decay σ (whereas the inverse Fourier transformation represents the function as a combination of undamped harmonics). Still if σ is negative, the direct transformation can be seen as the Fourier transformation of the function f(t) after correction of the latter for damping by multiplying the $e^{-\sigma t}$, at least if f(t) = 0 for t < 0. If the latter condition is fulfilled, the Laplace transformation for σ = 0 is identical to the Fourier transformation. The ratio of Laplace transforms $\underline{H}(s) \ = \ \underline{X}(s)/\underline{F}(s)$ is called the "transfer function", a concept which has found widespread use in signal analysis.

The Laplace transform of Equations 5 and 6, provided initial conditions (x and $\dot{x}$) are zero, leads to

$$(ms^2 \ + \ cs \ + \ k) \ \underline{X} \ = \ \underline{F}(s) \tag{40}$$

$$\underline{H}(s) \ = \ \frac{\underline{X}(s)}{\underline{F}(s)} \ = \ \frac{1/m}{s^2 \ + \ C/m \ s \ + \ k/m} \ = \ \frac{1/m}{s^2 \ + \ 2\zeta/\omega_n s \ + \ \omega_n^2} \tag{41}$$

$\underline{H}(s)$ for $s \ = \ j\omega$ is identical to the frequency response function, since the impulse response starts at t = 0.

The roots of the free vibration equation

$$ms^2 \ + \ Cs \ + \ k \ = \ 0 \tag{42}$$

are, for undamped systems,

$$s_{1,2} = -\sigma_d \pm j\omega_d \qquad \sigma_d = C/2m \tag{43}$$
$$(s_2 = s_1^\star) \qquad \omega_d = \sqrt{1 - \zeta^2}\,\omega_n$$

Using these roots, $\underline{H}(s)$ can be written

$$\underline{H}(s) = \frac{1/m}{(s - s_1)(s - s_1^\star)} \tag{44}$$

and by partial fraction expansion,

$$\underline{H}(s) = \frac{\underline{A}}{(s - s_1)} + \frac{\underline{A}^\star}{(s - s_1^\star)} \tag{45}$$

with

$$\underline{A} = \frac{1/m}{2j\,\omega_d}$$

the residue of the transfer function.

For $s = j\omega$ ($\sigma = 0$), this expression is identical to Equation 33, but it appears that for

$$s = s_1 = -\sigma_d + j\omega_d \quad \text{or} \quad s = s_2 = -\sigma_d - j\omega_d$$

$\underline{H}(s)$ becomes infinite, i.e., by premultiplying the damped impulse response with $e^{+\sigma dt}$ (hence correcting for damping), an undamped harmonic impulse response with frequency ω_d is obtained, the transform of which is identical to the really undamped frequency response, however, with resonance at $\omega = \omega_d$.

The advantage of the Laplace transform in comparison with the Fourier transform is purely mathematical; it allows the representation of the transfer function in terms of the complex roots (the "poles") of the damped free vibration equation, as in Equations 44 and 45. In the next section, the Laplace transform will merely be used as a mathematical tool for obtaining the frequency response function (i.e., the transfer function for $\sigma = 0$), which was defined above and which is measured in practice.

C. Dynamic Analysis of Multiple Degrees of Freedom (MDOF) Systems

1. Degrees of Freedom

For the simple mass-spring system of Figures 1 and 2, the time variation of one single parameter, x, completely describes motion. Such systems are SDOF systems. Multiple degrees of freedom systems (MDOF) can be built up by combinations of mass particles interconnected by springs and dashpots. Such "lumped parameter" systems (see Figure 8) are only interesting from the modeling point of view.

However, real structures as the beam of Figures 1 and 2 are "distributed parameter" systems: mass, stiffness, and damping characteristics are distributed continuously over the structure. Furthermore, these structures have an infinite number of degrees of freedom: in order to describe their motion fully, the motion of each differential mass element dm must be determined.

We consider the beam example. The free vibration Equation 8 must be set up for each mass element $dm = \mu\,dx$ (μ, mass per unit length). This results in a partial differential equation, since the position along the beam, x, as well as the time, t, must be taken as

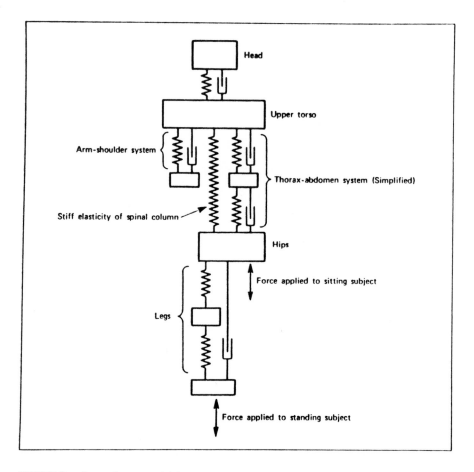

FIGURE 8.   Lumped mass model for the human body standing on a vertically vibrating platform. (From Broch, J. T., *Mechanical Vibration and Shock Measurements*, Bruël and Kjaer, Eds., Naerum, Denmark, 1980, 86. With permission.)

independent variables: the time variation of a function y(x), thus, y(x,t), rather than of one parameter x is the solution searched for. Whereas this partial differential equation can be solved analytically for a uniform beam, one can imagine that this is not feasible for more complex structures.

Therefore, below we will only refer to the analytical solution results for the beam vibration, whereas for the general theoretical discussion we will make use of a ''discretization procedure''. Discretization is a procedure by which a continuous function is replaced by a row of discrete points or variables. By discretization, a distributed parameter system is reduced to a MDOF system: the number of degrees of freedom is equal to the number of independent parameters. As an example, the beam of Figure 9 is discretized by taking the deflection y(t) of 13 equidistant points as characteristic independent parameters. As was the case for a SDOF system, the dynamic behavior of a MDOF system is intimately related to its free vibration characteristics.

2. Undamped Free Vibration of a MDOF System: Modal Parameters

In contradiction with the assumption made in Sections II.A and B, the deflection shape of the beam in the latter's free vibration is not unique, e.g., the shape represented in Figure 9 can be imagined as a free vibration shape as well. However, not any shape can be a free vibration shape, as is explained below.

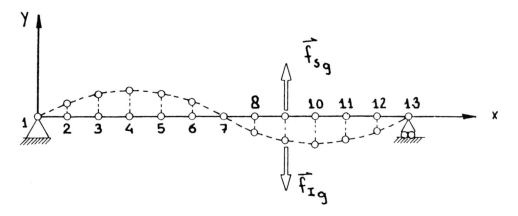

FIGURE 9.   Discretization: continuous beam represented as a 13° of freedom system.

We assume the broken line of Figure 9 to represent the beam's vibration shape in one of its two extreme positions: the plotted y_i values are the amplitudes Y_i in the equation

$$y_i = Y_i \sin \omega t \qquad (46)$$

ω is the common frequency of all harmonic oscillations of points i, which move in phase with one another (points 8 to 12 are shifted by 180° with respect to points 2 to 6).

The shape, as represented by the set of amplitudes Y_i, can only be a free vibration shape if, at the associated frequency ω, dynamic equilibrium between inertial and stiffness (elastic) forces exists in all points:

$$f_{I_i} + f_{S_i} = 0 \qquad (47)$$

Elastic and inertial forces in one point, e.g., 9, are not only determined by the displacement y_9, respectively, acceleration $\ddot{y}_9$, but also by the displacements, respectively, accelerations, in all other points.

$$f_{S_9} = -(k_{91} y_1 + \ldots + k_{99} y_9 \ldots + k_{9,11} y_{11}) \qquad (48)$$

$$f_{I_9} = -(m_{91} \ddot{y}_1 + \ldots + m_{99} \ddot{y}_1 \ldots + m_{9,11} \ddot{y}_{11}) \qquad (49)$$

The coefficients k_{ij} are the "stiffness influence coefficients"; the coefficients m_{ij} are the "mass influence coefficients". The set of dynamic equilibrium conditions in all n points is represented by a matrix equation, which, after substitution of the solution $y_i = Y_i \sin \omega t$ (see Equation 46) has the form

$$\left\{ \begin{bmatrix} k_{11} \text{----} k_{1n} \\ \vdots \quad \diagdown \quad \vdots \\ k_{n1} \text{----} k_{nn} \end{bmatrix} - \omega^2 \begin{bmatrix} m_{11} \text{----} m_{1n} \\ \vdots \quad \diagdown \quad \vdots \\ m_{n1} \text{----} m_{nn} \end{bmatrix} \right\} \begin{Bmatrix} Y_i \\ Y_n \end{Bmatrix} = 0 \qquad (50)$$

or symbolically

$$\{ \overline{K} - \omega^2 \, \overline{M} \} \, \overline{Y} = 0 \qquad (51)$$

$\overline{K}$ is the stiffness matrix, $\overline{M}$ is the mass matrix, and $\overline{Y}$ is the "mode shape" vector. $\overline{K}$ and $\overline{M}$ are symmetrical matrices (because of the reciprocity law $k_{ij} = k_{ji}$ and $m_{ij} = m_{ji}$).

From Equations 50 and 51, nonzero values for Y_i (the mode shape) can only be found if the determinant is zero:

$$|\overline{K} - \omega^2 \overline{M}| = 0 \tag{52}$$

Equation 52 is the "characteristic equation". The solutions ω_k of this characteristic equation are the natural frequencies.

The number of natural frequencies is equal to n, the number of degrees of freedom. For each value of ω_k, the associated mode shape (within a proportionality constant) can be calculated from Equations 50 and 51.

The free vibration of a MDOF system consists of a combination of n modes, each mode having its own shape and its own natural frequency. Analytical solution of the continuous beam case leads to exactly the same conclusion: the number of modes, however, is infinite. In practical applications, however, we are only interested in a limited frequency range and thus in a limited number of modes: the ones occurring within this frequency range. Mode shapes of a uniform slender beam, deforming in pure bending (shear deformation and rotatory inertia are neglected), under different support conditions ("boundary conditions"), are represented in Figure 10. In this case, there is a very simple formula for the associated natural frequencies (ω_k, circular frequency; f_k, cyclic frequency):

$$\omega_k = 2\pi f_k = 2\pi\alpha \sqrt{\frac{EI}{\mu\ell^4}}$$

E, Young's modulus
I, second moment of area of the beam cross section $\left.\right\}$ EI, bending stiffness   (53)
μ, mass per unit length
ℓ, beam length

The proportionality constant α depends upon mode shape and boundary conditions; values are indicated in the figure. Referring to Section II.B.1,

$$\omega_k = \sqrt{\frac{K_k}{M_k}} \tag{54}$$

Generalized stiffness K_k is proportional to the bending stiffness EI, generalized mass M_k is proportional to $\mu\ell^4$, and both depend upon the mode shape (α), the latter depending upon support conditions.

The modes shapes, $\{Y\}_k$, and the associated natural frequencies, ω_k, are called the "modal parameters" of the undamped free vibration. Since the modal parameters are determined by the structure's spatial mass and stiffness properties (see Equations 50 and 51), it is in principle possible to determine the latter from the former. In other words, the modal parameters depict a "mechanical image" of the structure. For reference, if a human bone could be considered as a uniform beam vibrating in bending, its bending stiffness could be obtained from natural frequency measurements using Equation 53.

3. Relation Between Free Vibration Modal Parameters and Dynamic Response of MDOF Systems

In a continuous system, mass and stiffness as well as damping characteristics are distributed parameters, which can be discretized. The same is true in general for the external forces. For example, a bone in vivo is subject to muscle, ligament, and joint reaction forces (and contacting forces with surrounding tissues) which are distributed over areas, rather than applied in discrete points. However, one specific case is of special interest, namely, the

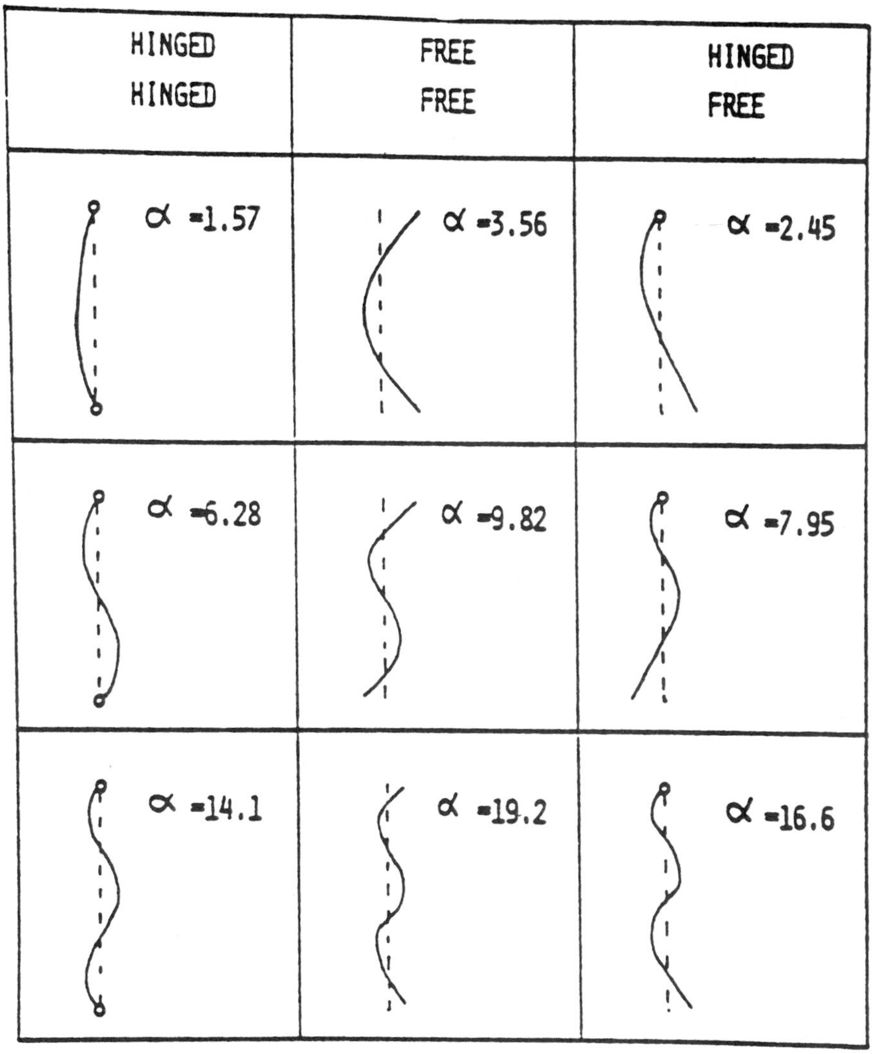

HINGED HINGED	FREE FREE	HINGED FREE
$\alpha = 1.57$	$\alpha = 3.56$	$\alpha = 2.45$
$\alpha = 6.28$	$\alpha = 9.82$	$\alpha = 7.95$
$\alpha = 14.1$	$\alpha = 19.2$	$\alpha = 16.6$

FIGURE 10.   Bending mode shapes of a simple beam in three standard support conditions; α is the proportionality coefficient in the formula for the cyclic natural frequency f_o

$$f_o = \alpha \sqrt{\frac{EI}{\mu \ell^4}}$$

where EI, bending stiffness; μ, mass per unit length; and ℓ, length.

bone's response to an impulse exerted in one discrete point. This case reflects the situation of vibration testing, where the impulse response is measured with the aim of analyzing the system's free vibration characteristics.

The dynamic response of a system to an arbitrary force distribution is the sum of its responses to the forces exerted in all points (superposition principle). Hence, general dynamic response is directly related to impulse response characteristics. In Sections II.C.3.a and b, we will treat the single point impulse response characteristics which are fundamental in vibration testing, whereas in Section II.C.3.c, we will briefly deal with the question how the response to actual dynamic loading can be calculated.

a. Point Impulse Response and Frequency Characteristics of MDOF Systems

If a MDOF structure is given a unit impulse in a point j, it responds by free vibration movement, just like a SDOF structure. However, now the free vibration in all points i (i = 1,n) must be considered, and furthermore, vibration is (or can be) now a combination of a number of vibration modes. Let us consider for one moment the (undamped) case where only one vibration mode is excited. Then, all points, i, will describe harmonic oscillations, the amplitudes of which are the components Y_i of the modal vector. This example illustrates the fact (to be generalized further) that mode shapes can be determined from impulse response characteristics.

The displacement response $h_{ij}(t)$ of point i to an impulse exerted in point j allows the calculation of the response $x_{ij}(t)$ to an arbitrary force $f_j(t)$ by convolution:

$$x_{ij}(t) = h_{ij} \ast f_j \tag{55}$$

and hence in the frequency domain,

$$\underline{X}_{ij}(\omega) = \underline{H}_{ij}(\omega) \, x\underline{F}_j(\omega) \tag{56}$$

Now, if forces are exerted in all n points, the resulting displacement is

$$x_i(t) = \sum_{j=1,n} x_{ij}(t) = \sum_{j=1,n} h_{ij}(t) \ast f_{(j)} \tag{57}$$

and in the frequency domain,

$$\underline{X}_i(\omega) = \sum_{j=1,n} \underline{H}_{ij}(\omega) \times \underline{F}_j(\omega) \tag{58}$$

The latter relation, valid for all values of i, can be represented in matrix notation

$$\overline{\mathbf{X}}(\omega) = \overline{\mathbf{H}}(\omega) \cdot \overline{\mathbf{F}}(\omega) \tag{59}$$

$\overline{\mathbf{H}}(\omega)$ is the frequency response matrix. In Laplace notation,

$$\overline{\mathbf{X}}(s) = \overline{\mathbf{H}}(s) \cdot \overline{\mathbf{F}}(s) \tag{60}$$

$\overline{\mathbf{H}}(s)$ is the transfer matrix. It is identical to the frequency response matrix for $s = j\omega$ ($\sigma = 0$).

For $h_{ij}(t)$, the reciprocity law holds:

$$h_{ij}(t) = h_{ji}(t) \tag{61}$$

The same is true for the frequency response functions and transfer functions:

$$\underline{H}_{ij}(\omega) = \underline{H}_{ji}(\omega) \tag{62}$$

$$\underline{H}_{ij}(s) = \underline{H}_{ji}(s)$$

Thus, the transfer matrix is symmetrical.

The undamped free vibration modal parameters of a structure are determined by the mass and stiffness matrices (see Equations 50 and 51). We will now look how the transfer matrix depends upon the mass, stiffness, and damping matrices. At the same time we will introduce the effects of damping upon the free vibration modal parameters. This will allow us to quantify the relationship between transfer matrix and free vibration modal parameters.

The dynamic equilibrium equation holds in each point i:

$$f_{I_i} + f_{S_i} + f_{D_i} + f_i(t) = 0 \qquad (63)$$

The inertial and stiffness forces stem from Equation 46, while the damping force,

$$f_{D_i} = -\sum_{j=1,n} C_{ij} \dot{x}_j \qquad (64)$$

in which C_{ij} are the damping influence coefficients. The dynamic equilibrium Equation 63 is written for the whole system in terms of the mass, damping, and stiffness matrices and the displacement and force vectors:

$$[M]\{\ddot{x}\} + [C]\{\dot{x}\} + [K]\{x\} = \{f(t)\} \qquad (65)$$

which, after Laplace transformation, becomes:

$$(\overline{\mathbf{M}}^2 + \overline{\mathbf{C}}s + \overline{\mathbf{K}})\,\overline{\mathbf{X}}(s) = \overline{\mathbf{F}}(s)$$
$$\overline{\mathbf{B}}(s)\,\overline{\mathbf{X}}(s) = \overline{\mathbf{F}}(s) \qquad (66)$$

and thus

$$\overline{\mathbf{H}}(s) = \overline{\mathbf{B}}(s)^{-1} \qquad (67)$$

$\overline{\mathbf{H}}(s)$ is the inverse matrix of $\overline{\mathbf{B}}(s)$, the ''dynamic stiffness matrix''.

$$\overline{\mathbf{H}}(s) = \frac{\mathrm{Adj} \cdot [\overline{\mathbf{B}}(s)]}{\det [\overline{\mathbf{B}}(s)]} \qquad (68)$$

Adj·$[\overline{\mathbf{B}}(s)]$ means adjoint matrix of $\overline{\mathbf{B}}(s)$; det $[\overline{\mathbf{B}}(s)] = 0$ is the characteristic equation for the free vibration, which is equivalent to Equation 52 if $\overline{\mathbf{C}} = 0$ (undamped).

When damping is present, the solutions of the characteristic equation

$$\det [\overline{\mathbf{B}}(s)] = 0 \qquad (69)$$

are complex conjugate pairs:

$$s_k = \sigma_k \pm j\omega d_k \qquad (70)$$

σ_k is the exponential decay rate, ω_{d_k} is the damped natural frequency of mode k.

As we did for the damped SDOF system, we write the value of det $[\overline{\mathbf{B}}]$ in function of these solutions (see Equation 70) (the poles), which, after partial fraction expansion, results in the following expression for the transfer function $H_{ij}(s)$:

$$H_{ij}(s) = \sum_{k=1}^{n} \frac{\mathbf{A}_{ij}^k}{(s - s_k)} + \frac{\mathbf{A}_{ij}^{k\star}}{(s - s_k\star)} \qquad (71)$$

The frequency response function is the same expression for $s = j\omega$. We conclude that the

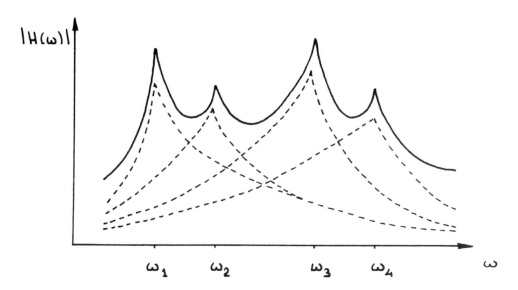

FIGURE 11.   Frequency response of n-DOF systems (full line) and superposition of n-SDOF frequency response functions (broken lines).

frequency response function of a MDOF system is the sum of k frequency response functions of the type (see Equation 33 and Figure 11). The parameter $\underline{A}_{ij}^{k}$ is the amplitude (complex, thus including the phase) of the kth damped free vibration mode in the point i under the influence of a unit impulse exerted in j. Hence the parameters $\underline{A}_{ij}^{k}$, $\underline{A}_{ij}^{k\star}$ depict the mode shapes. For $s = j\omega$ and introducing

$$\underline{A}_{ij}^{k} = U_{ij}^{k} + jV_{ij}^{k}$$

$$s_{k} = \sigma_{k} + j\omega d_{k}$$

Equation 71 becomes:

$$H_{ij}(\omega) = \sum_{k=1}^{n} \left(\frac{U_{ij}^{k} + jV_{ij}^{k}}{j(\omega - \omega_{d_{k}}) - \sigma_{k}} + \frac{U_{ij}^{k} - jV_{ij}^{k}}{j(\omega + \omega_{d_{k}}) - \sigma_{k}} \right) \tag{72}$$

From Equations 71 and 72, it appears that mode shapes, damped resonant frequencies $\omega_{d_{k}}$, and damping values ($\sigma_{k} = \omega_{k} \times \zeta_{k}$) (the modal parameters) can be derived from the set of frequency response functions in all points i (i = 1,n) to excitation in point j, hence from one column of the the transfer matrix. Since the latter is symmetrical, the same holds for one row of the matrix, i.e., the frequency response in point j to excitation in all points i. The procedure in which modal parameters are derived from experimentally measured frequency response functions is called "modal analysis". Mode shapes of a dry tibia in the "free-free" condition obtained from experimentally measured frequency response functions are shown in Figure 28.

b. Influence of Damping — Proportional and Nonproportional Damping

The parameters $U_{ij}^{k} \pm V_{ij}^{k}$ depict the damped mode shapes. In general when damping is present, the phase angle

$$\arctan \frac{V_{ij}^{k}}{U_{ij}^{k}} \tag{73}$$

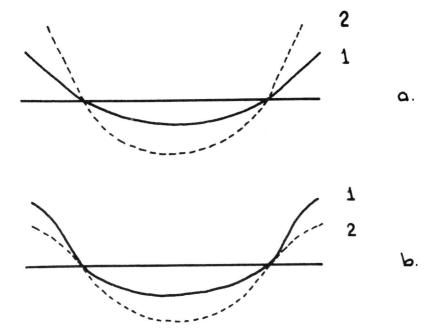

FIGURE 12.   Effect of nonproportional damping on mode shapes. (a) Real mode (undamped, proportionally damped); (b) complex mode, "fluttering" (unproportionally damped).

is different from one point to another, hence, all points are not oscillating in phase with each other. Vibration proceeds by traveling waves instead of standing waves (see Figure 12). Mode shapes are complex (represented by complex numbers). It can be shown, however, that if

$$\overline{C} = a\overline{M} + b\overline{K} \qquad (74)$$

in which a and b are arbitrary scalars, the so-called "proportional damping" case, the damped mode shapes are indentical to the undamped ones and thus represent "standing waves". They are called "real modes" as they can be represented by real numbers A_{ij}

$$U_{ij}^k + {}_jV_{ij}^k = A_{ij}\, e^{j\varphi k} \qquad (75)$$

φ_k being equal in all points.

There is a typical difference in the variation of the modal damping coefficients ζ_k with ω_k in the case where damping is proportional to mass (b = 0) and the case where damping is proportional to stiffness (a = 0). As we already mentioned in connection with the beam vibration (see Section I.A), an individual vibration mode of a structure can be modeled as an individual SDOF mass-spring-dashpot system:

$$m_k s^2 + c_k s + k_k = 0 \qquad (76)$$

in which m_k, c_k, and k_k are, respectively, the generalized mass, generalized damping coefficient, and generalized stiffness. As we will see in the paragraph on modal coordinates (corresponding to these generalized parameters), $\overline{C} = a\overline{M}$ implies $c_k = am_k$ and $\overline{C} = b\overline{K}$ implies $c_k = bk_k$. The damping ratio $\zeta_k = [C/(2\sqrt{km})]_k$, and ω_k (undamped) = $(\sqrt{k/m})_k$; hence, if $c_k = am_k$: $\zeta_k = (a/2)(1/\omega_k)$; the damping ratio is inversely proportional to ω_k; if $c_k = bk_k$: $\zeta_k = b_b \times \omega_k$, the damping ratio is proportional to ω_k.

c. Response to Actual Dynamic Loading: Modal Coordinates

The SDOF model free vibration equation,

$$m_k s^2 + c_k s + k_k = 0 \tag{77}$$

is in its full version

$$(m_k s^2 + c_k s + k_k)\,\underline{X}_k(s) = 0$$

which is the Laplace transform of

$$m_k \ddot{x}_k + c_k \dot{x}_k + k_k x_k = 0 \tag{78}$$

in which x_k is one single parameter determining the motion completely, e.g., the deflection of the middle point of the beam (see Figure 1B). Indeed, if the mode shape ψ_k is known, the amplitude in one point completely defines all other amplitudes: x_k is called a "modal coordinate".

Hence, if all n mode shapes ψ_k are known, the matrix free vibration Equations 50 and 51 can be written as a set of n independent equations (see Equation 77). By introducing the parameters m_k, c_k, k_k, and the modal coordinates, the n DOF system is uncoupled into n SDOF systems.

It would be very interesting if this uncoupling procedure could be extended to the general dynamic response problem, i.e., if the matrix equation

$$[M]\{\ddot{y}\} + [C]\{\dot{y}\} + [K]\{y\} = \{f(t)\} \tag{79}$$

(in which $\{y\}$ is the displacement vector: $\{y_1 \ldots y_i \ldots y_n\}$ and $\{f(t)\}$ is the external force "vector" $\{f_1(t) \ldots f_n(t)\}$ and its Laplace transform

$$(\overline{\mathbf{M}}s^2 + \overline{\mathbf{C}}s + \overline{\mathbf{K}})\,\overline{\mathbf{Y}} = \overline{\mathbf{F}}(s) \tag{80}$$

could be reduced to a set of n independent equations

$$m_k \ddot{y}_k + c_n \dot{y}_k + k_k = f_k(t) \tag{81}$$

or

$$(m_k s^2 + c_k s + k_k)\,\underline{Y}_k(s) = \underline{F}_k(s) \tag{82}$$

in which $f_j(t)$ is the "generalized force" and $\underline{F}_k(s)$ is the Laplace transform of the latter.

Indeed, this transformation is possible. Since the motion $\{y(t)\}$ [i.e., $y_1(t) \ldots y_i(t) \ldots y_n(t)$] is a combination of free vibration modes, it can be written:

$$\{y\} = \Sigma_k \{\psi_k\}\, Y_k(t) \tag{83}$$

in which the mode shape is represented by vector $\{\psi_k\}$, and a row of values $\{\psi_1 \ldots \psi_i \ldots \psi_n\}_k$ and $Y_k(t)$ is a time varying scalar, associated to mode k, i.e., the "coordinate" which, multiplied by the mode shape, gives the actual deflections in all points. This expression can be rewritten:

$$\{y(t)\} = [\psi]\{y(t)\} \tag{84}$$

in which $[\psi]$ is the modal matrix $[\{\psi_k\} \ \ldots \ \{\psi_k\}]$. With this notation, the undamped dynamic response equation

$$[M]\{\ddot{y}\} + [K]\{y\} = \{f(t)\} \tag{85}$$

can be written:

$$[M][\psi]\{\ddot{y}(t)\} + [K][\psi]\{y(t)\} = \{f(t)\} \tag{86}$$

Now, the following properties exist, for any two mode shapes $[\psi]_\ell$ and $[\psi]_k$:

$$[\psi]_\ell^t[M][\psi]_k = 0 \tag{87}$$

$$[\psi]_\ell^t[k][\psi]_k = 0 \tag{88}$$

($[\psi]_\ell^t$ means the transpose of $[\psi]_\ell$), the transpose is formed by interchanging rows and columns in the matrix, and furthermore

$$[\psi]_k^t[M][\psi]_k = m_k, \ \text{(a scalar)} \tag{89}$$

$$[\psi]_k^t[K][\psi]_h = k_k \quad \text{and} \quad k_k = \omega_k^2 m_k \tag{90}$$

These properties (see Equations 87 and 88) are known as the "orthogonality properties", the mode shapes are "orthogonal" with respect to the mass and stiffness matrices.

Premultiplying Equation 86 by $[\psi]_k^t$ and applying Equations 87, 88, 89, and 90 yields the undamped SDOF equilibrium equation we look for:

$$m_k\ddot{y}_k + k_ky_k = f_k(t) \tag{91}$$

in which $f_k(t)$, the generalized force, is

$$f_k(t) = [\psi]_k^t f(t) \tag{92}$$

y_k is the modal coordinate of mode k. Since the mode shapes $\{\psi_k\}$ are given within a scalar coefficient, they can be determined such that $[\psi]_k^t[M][\psi]_k = 1$. $\{\psi_k\}$ vectors obeying this relation are "normalized mode shapes" and the associated model coordinates y_k are "normal coordinates".

Obviously, the Laplace (frequency domain) version of Equation 91 is

$$(m_ks^2 + k_k) \ \underline{Y}_k(s) = \underline{F}_k(s) \tag{93}$$

In the case of proportional damping, the damping matrix obeys to properties similar to Equations 87, 88, 89, and 90 and the mode shapes are unaltered, hence Equations 81 and 82 are obtained the same way as Equations 91 and 93, with

$$C_k = [\psi]_k^t[C][\psi]_k \tag{94}$$

In the nonproportionally damped case, the theory still holds if the complex damped mode shapes are used. Application becomes complicated.

As a conclusion, we can say that the dynamic response to an arbitrary "force vector" $\{f(t)\}$ can be calculated as follows:

1.   Determination of free vibration modal parameters
2.   Decoupling of the equation of motion, by premultiplying with $[\psi]_k^t$, hence determination of m_k, c_k, k_k, and $f_k(t)$ in Equation 81
3.   Solution of Equation 81, e.g., via the frequency domain
4.   From the solutions $\{Y_k(t)\}$, the real displacement vector $\{y(t)\}$ is obtained by

$$\{y(t)\} = \Sigma_k \{\psi_k\} \, Y_k(t) = [\psi][Y] \qquad (95)$$

This is the "modal superposition method", which is purely computational and requires the mass, stiffness, and damping matrices to be known. Alternative procedures are based upon the direct frequency response superposition concept (transfer matrix, Equation 59) and can make use of experimentally determined frequency response data.

4. Discretization and Modeling

    The simplest MDOF model for a continuous structure is the "lumped parameter" model, e.g., Figure 8. Several physical models, combining lumped parameter elements with beams, have been set up for specific cases. In general, however, one can distinguish between three types of more advanced mathematical models:

1.   The model representing the structure by a number of n discrete points, which we have used thus far (see Figure 9). It has to be noted that, in general, instead of one single displacement coordinate x_i, three-dimensional displacement coordinates (hence 3n degrees of freedom) are needed. One of the major problems in such models is the evaluation of the structural property (mass, stiffness, and damping) matrices. As far as mass is concerned, both the lumped mass (assigning a portion of total mass to each point) and the more advanced consistent mass matrix approach (determining mass influence coefficients from their definition using interpolation functions as shown in No. 3) are in use.
2.   The "generalized coordinate" model, representing the shape of the structure as a combination of n known functions, e.g., for a linear beam

$$y(x) = \Sigma_n Z_n \psi_n(x) \qquad (96)$$

    Z_n are the general coordinates. The above-mentioned "modal coordinates" belong to this category.
3.   The finite element model, which is a combination of No. 1 and 3. The structure is divided into a number of block elements. The connecting points between block elements are the nodal points, the displacements (including beam rotations) of which are taken as generalized coordinates. The deflection of the complete structure is expressed in terms of these coordinates using interpolation functions, which define the shape between the specified nodal displacements.

D. Frequency Response Testing

    A typical frequency response test involves:

1.   Exerting a time varying force in one point of the structure, e.g., by an electromagnetic shaker (Figure 13) or a hammer (Figure 14). In the early experiments, steady-state response measurements were made by feeding a harmonic input signal to the shaker and varying ω step by step. Nowadays, using the Fast Fourier Transform (FFT) algorithm in digital data processing, signals containing a large number of harmonics are used (random, periodical random, swept sine, etc., Figure 15), allowing quick

FIGURE 13A.   Electromagnetic shaker Bruël and Kjaer type 4809. (From Broch, J. T., *Sectional Drawing of Vibration Exciter Type 4809,* Bruël and Kjaer, Eds., Naerum, Denmark, 1981, 217. With permission.)

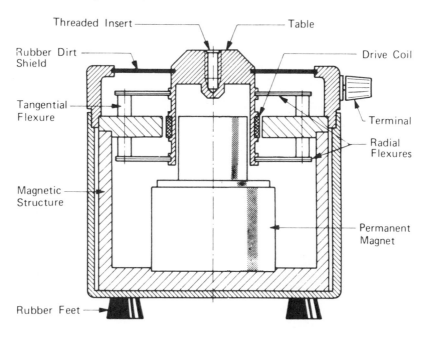

FIGURE 13B.   Sectional drawing.

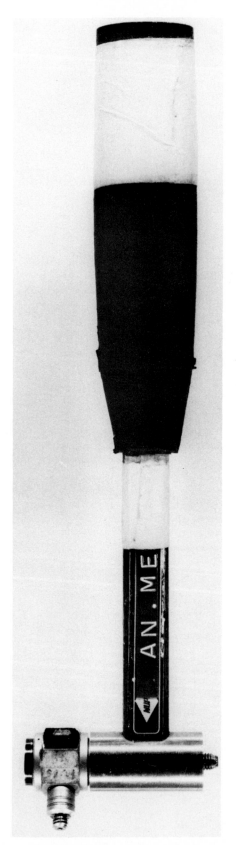

FIGURE 14.   Instrumented hammer (with force transducer PCB-208).

TIME   FREQUENCY

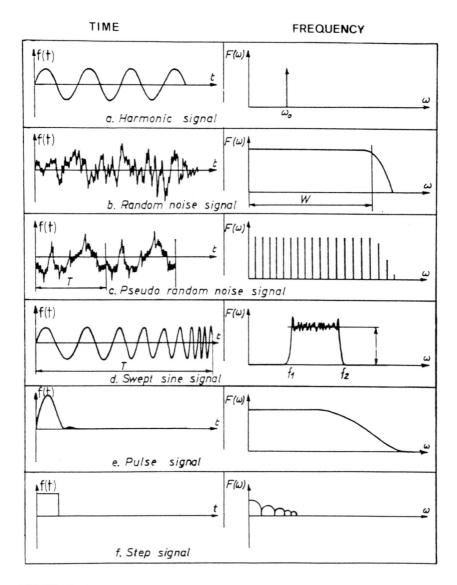

FIGURE 15. Excitation signals and their frequency content. (From Péters, J., *Proc. Int. Seminar Modal Analysis,* Van Brussel, H. and Snoeys, R., Eds., Katholieke Universiteit Leuven, 1977, 18. With permission.)

measurements and averaging over a number of tests. The impact signal from a hammer also spreads over a large frequency range (Figure 16).

2. Measuring the resulting displacement in one (the excitation point or another point) or several points

In dynamic measurements, generally the acceleration of a point is measured instead of the displacement, using an accelerometer. An accelerometer is an electromechanical transducer which produces at its output terminals a voltage or charge that is proportional to the acceleration to which it is subjected.

In the accelerometer, a piezoelectric element is loaded by a mass and a preloading spring or ring. When subjected to vibration, the mass exerts a varying force on the piezoelectric element which is directly proportional to the vibratory acceleration. For frequencies lying

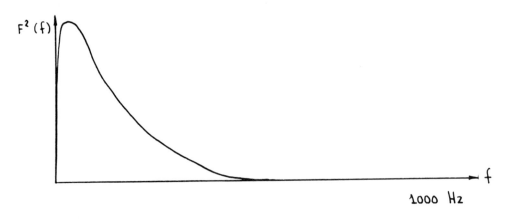

FIGURE 16.   Frequency spectrum of a hammer impact on a tibia in vivo. $F^2(f)$, squared force amplitude.

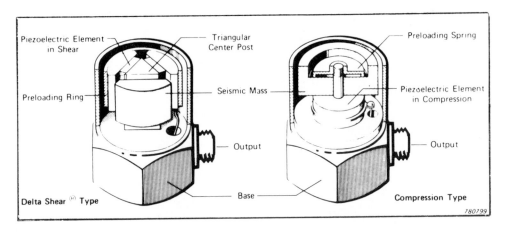

FIGURE 17.   Two accelerometer configurations in common use. (From Broch, J. T., *Mechanical Vibration and Shock Measurements,* Bruël and Kjaer, Eds., Naerum, Denmark, 1981, 100. With permission.)

well under the resonant frequency of the assembly, the acceleration of the mass will be the same as the acceleration of the base, and the output signal level will be proportional to the acceleration to which the accelerometer is subjected (Figure 17). From the acceleration, the velocity and displacement can be obtained by integration in the time domain or by dividing by ω (combined with phase shift $\pi/2$), respectively, by ω^2 (combined with phase shift π) in the frequency domain.

In practical accelerometer designs, charge preamplifiers are used to make the signal suitable for connection to the measuring and analyzing instrumentation. Like the accelerometer, the force transducer also uses a piezoelectric element which, when compressed, produces an electrical output proportional to the force transmitted throught it (Figure 18).

An impedance head contains two transducers, a force transducer and an accelerometer (Figure 19). Normally the output of the accelerometer is integrated to obtain a signal proportional to velocity, so that the mechanical impedance $\underline{Z}(\omega) = \underline{F}(\omega)/\underline{V}(\omega)$ can be determined.

A vibration test can be run (1) only monitoring the acceleration signal (''resonance testing'') and (2) monitoring force as well as acceleration. In a typical digital data processing setup, the signal is amplified, low-pass filtered, ''sampled'', i.e., transformed in a set of discrete numbers taken at a constant time interval (in an analog to digital converter, ADC), and Fourier transformed using a FFT algorithm (Figure 20). For the details of digital data processing, we refer to the relevant literature.[4]

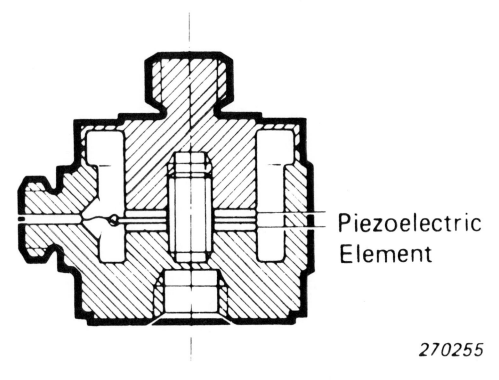

270255

FIGURE 18.   Construction of a typical force transducer (Bruël and Kjaer 8200). (From Broch, J. T., *Mechanical Vibration and Shock Measurements*, Bruël and Kjaer, Eds., Naerum, Denmark, 1981, 120. With permission.)

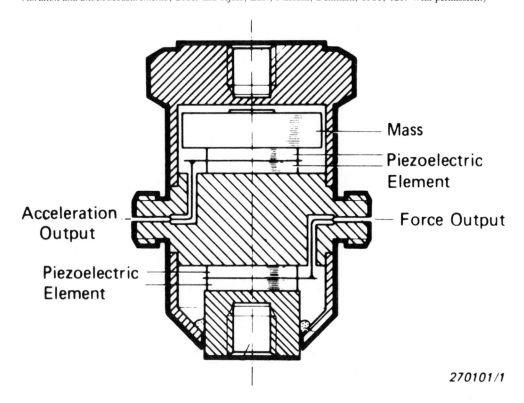

270101/1

FIGURE 19.   Construction of a typical impedance head (Bruël and Kjaer 8001). (From Broch, J. T., *Mechanical Vibration and Shock Measurements*, Bruël and Kjaer, Eds., Naerum, Denmark, 1981, 121. With permission.)

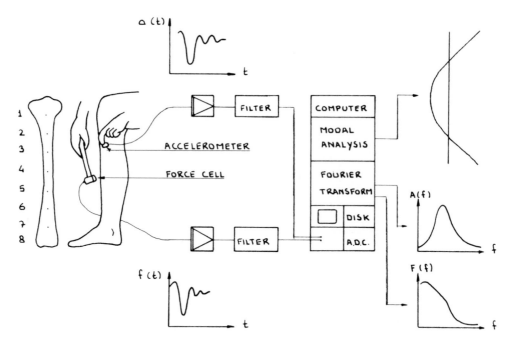

FIGURE 20.   Digital data processing and modal analysis setup. Force signal f(t) and acceleration signal a(t) are amplified, low-pass filtered, converted into digital signals (ADC, analog to digital converter) and Fourier-transformed into F(f) and A(f), respectively. The computer allows the computation of the frequency response function and other relevant functions (autopower and cross power spectra, etc.). Modal analysis software allows the computation of modal parameters.

In experimental frequency-response function measurements, Fourier-transformed signals of acceleration and force are divided by each other. The procedure involves averaging over a number of measurements. Alternative, e.g., analog, procedures for the determination of resonant frequencies are in use.

III. IN VIVO TESTING OF LONG BONES

The human musculoskeletal system can be considered as a mechanical structure of a very particular nature, as its properties are in permanent evolution. Bones (muscles and ligaments) develop in a way which is strongly dependent upon the use made of them. As an effect of age, osteoporosis, i.e., a metabolic disease of bone tissue resulting in a decrease in strength and fracture toughness, occurs to a higher or lesser degree.

Furthermore, bone repair operations are made in several clinical situations, e.g., the natural biological repair process in fracture healing, the insertion of metallic fixation plates or nails, and the implantation of joint prostheses. Implantation of foreign load-bearing material leads to adaptive bone growth mechanisms.

Apparently, in orthopedic clinical practice such as in mechanical technology, there is a need for nondestructive techniques for quality control and monitoring. X-ray techniques are in widespread clinical use. Analysis of (ultrasonic) sound wave propagation in bones and dynamic response analysis of bone are still in the development stage.

The potentialities of dynamic response analysis in this area are investigated by a number of research groups, experimentally as well as theoretically. The whole problem can be summarized as follows: to establish a practical and reliable relation between the frequency response characteristics (e.g., natural frequencies) measured in well-defined experimental conditions on the one hand and the physical bone characteristics on the other hand.

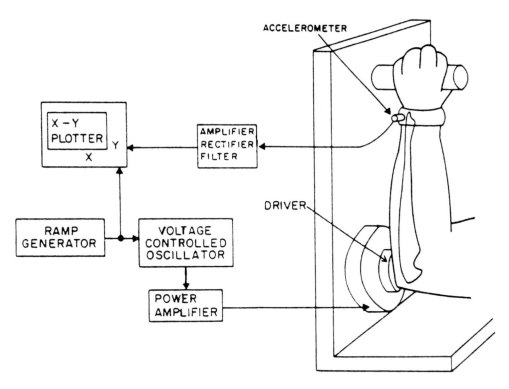

FIGURE 21.   Setup used by Jurist for ulnar resonant frequency measurement. (From Jurist, J. M., Hoeksema, H. D., Blacketter, D. A., Snider, R. K., and Garner, E. R., *Orthopaedic Mechanics, Procedures and Devices*, Vol. 3, Ghista, D. W. and Roaf, R., Eds., Academic Press, London, 1981, 4. With permission.)

For reference, we consider again the simple beam (see Figure 10). The characteristics of the structure itself (geometry, distribution of mass, and stiffness) as well as the boundary conditions are well defined. Hence it was simple to calculate the bending mode shapes and to establish for each mode the relation between the natural frequency on the one hand and flexural rigidity (EI), mass per unit length (μ), and length (ℓ) on the other hand;

$$f_k = \alpha \sqrt{\frac{EI}{\mu \ell^4}} \qquad\qquad (97)$$

In fact, the first models proposed for in vivo long bones were simply supported uniform beams.

A. Steady-State Vibration Tests In Vivo — Beam Models

The early investigations in the field were steady state frequency response tests, i.e., tests using a harmonic force input signal, with increasing values of the angular frequency ω. The dynamic models used were essentially SDOF elastic or viscoelastic beams.

1. Resonant Frequency Determinations

Jurist,[5] using the measuring setup of Figure 21, determined the resonant frequencies of human ulnas by varying the frequency of the harmonic input signal to a modified loudspeaker, used as a shaker, and measuring the bone response by an accelerometer. The reproducibility of resonant frequency measurements using this setup in varying test conditions was discussed by Jurist and Dymond.[6] Jurist[5] associated the first resonancy observed (at ±240 Hz) with the single bending mode. The ulna was modeled as a simple beam in the hinged-hinged condition. The frequency values obtained on a number of human subjects were claimed to correlate significantly with the degree of osteoporosis (see Section III.D). These results were questioned by Doherty et al.,[7] who suspected the obtained resonant frequencies to be associated with rigid body rather than bending modes and doubted about the assumed physical relation between the resonant frequency and the degree of osteoporosis (see Section III.D). The same authors[7] determined resonant frequencies of an excised tibia mounted in the hinged-hinged condition and stated all in vivo measuring attempts should be preceded by identification studies on excised bones.

Jurist and Kianian,[8] modeling the human ulna as an isotropic cylindrical tube, studied the sensitivity of the first natural frequency to the boundary conditions and to E, μ, and ℓ by analytical beam calculations. Three boundary models were compared:

1. Simply supported in the anterior-posterior plane
2. Supported by springs at both ends (representing wrist and elbow stiffness)
3. As above plus an additional mass-spring unit at the wrist, modeling the interposing skin and the accelerometer

Spiegl and Jurist[9] improved these models by including the mass of the bone marrow and taking a rigid body mode into account. The models were evaluated with data collected on 118 school children[8] and on an expanded population of 210 subjects aged 6 to 91 years.[9] These experimental data were compared with the resonant frequencies calculated from the models, using a nominal value for E (taken from literature) values for μ and I calculated from ulnar mineral content and width measured by the photon absorptiometric method,[10] ℓ values measured on the test subjects, and various estimated values for the spring constants representing joint stiffness.

The main conclusions were

1. The calculated rigid body frequencies were far below the experimentally measured resonances (in contradiction to the suggestion by Doherty et al.).[7]
2. Since Models II and III were virtually independent of the values for the springs (representing the wrist and elbow stiffness) over a wide range of spring values and since Model III was virtually independent of the accelerometer — skin parameters, the simplest Model I, including the bone marrow mass, was appropriate. Linear regressions between the resonant frequencies predicted from the latter model and the observed frequencies yielded a correlation coefficient R = 0.78 (Figure 22).[11]

It must be noted that the effect of vibrating muscle mass was not included at all in the modeling.

2. Driving Point Impedance Tests

Campbell and Jurist[12] studied the feasibility of determining the degree of union of healing femoral neck fractures by measurement of the driving point mechanical impedance on excised femora. Thompson[13] made driving point impedance measurements on the ulna in vivo. The impedance head was driven by an electromagnetic shaker and pressed against the ulna using a balance-counterweight system (Figure 23). The arm was placed in supports and plaster pads were used at the bony prominences of wrist and elbow. Using harmonic excitation with increasing ω, steady state impedance graphs were obtained as represented in Figure 24.

Orne[14] proposed a simply supported linear uniform viscoelastic beam as a model for Thompson's measurements (Figure 25). The skin interposing ulna and impedance head was modeled as a viscous solid, with material constants which vary with the static component

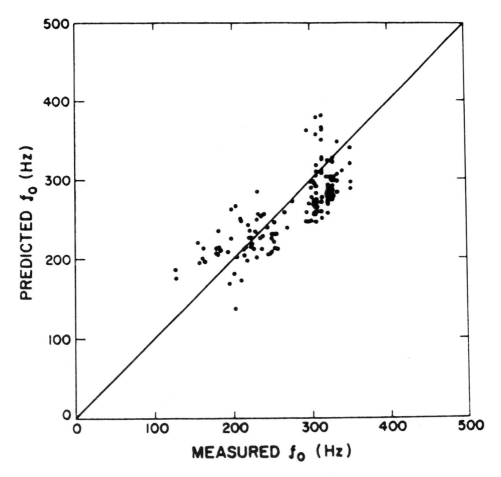

FIGURE 22. Observed and predicted ulnar resonant frequency obtained with the simply supported beam model, corrected for mass loading by bone marrow. (From Jurist, J. M., Hoeksema, H. D., Blackletter, D. A., Snider, R. K., and Garner, E. R., *Orthopaedic Mechanics, Procedures and Devices,* Vol. 3, Ghista, D. W. and Roaf, R., Eds., Academic Press, London, 1981, 18. With permission.)

of the forcing function. In the quasistatic region of the impedance graph, i.e., at very low subresonant frequencies, the absolute input impedance |Z| can be related to the combined stiffness only (neglecting mass and damping effects) of bone and skin. However, in order to determine the bone bending stiffness EI separately, the static preloads should be increased to levels far beyond the pain threshold, or independent experimental procedures must be developed to measure the skin stiffness accurately at various levels of preload. In connection with the interposing soft tissue effect, Saha and Lakes[16] made impulse response measurements of human tibia in vivo as well as on excised femurs with interposing rubber pads. Acceleration and force signals where monitored in the time domain. No saturation of soft tissue related effects could be observed at any preload level below the pain threshold.

Orne and Mandke[15] modified Orne's model to include the resistance to lateral vibration which is provided by the surrounding musculature. The musculature is represented as a continuous series of damped oscillators attached to the ulna and characterized by mass, spring, and damping coefficients, m, k_M, and c_M, per unit length of the ulna (Figure 26).

Using "physically plausible" values for the musculature parameters, the theoretical imped-ance curves could be made to fit Thompson's experimental results better than the previous model over a wide frequency range. Considerable improvement is achieved in the frequency range of 100 to 300 Hz when the muscle mass is accounted for (see Figure 24).

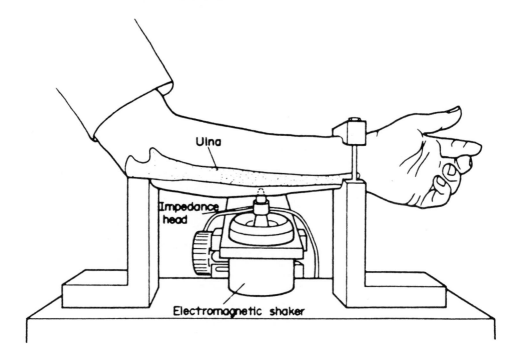

FIGURE 23.   Setup used by Thompson for the determination of ulnar impedance. (From Orne, D., *J. Biomech.*, 7, 250, 1974. With permission.)

As Orne and Mandke suggest, the model should be evaluated by measurements on in vivo ulnas, measurements with the test probe applied directly to the bone, and measurements on excised ulnas. Measurements of this kind were made on tibias by Van der Perre et al.[17] (see Section III.B).

A series of theoretical curves were generated to evaluate the sensitivity of ulnar impedance to changes in its modulus of elasticity, density, and cross-sectional dimensions. The impedance response curve is much more sensitive to changes in cross-sectional dimension than it is to changes in Young's modulus or density. Finally, it was suggested that a tapered instead of a uniform beam model for the ulna would be more appropriate.

As a conclusion we can say that while impedance curves, when interpreted using an appropriate mathematical model, allow theoretically the determination of stiffness, mass, and damping parameters separately, on the other hand they are very sensitive to a number of disturbing influences: skin compliance (static preload), location of point of application, and muscle parameters.

B. Generalized MDOF Dynamic Analysis of Long Bones

In Section III.A we discussed steady-state vibration response tests in which the boundary conditions were assumed (based on the experimental setup used, on estimates of joint elastic properties, and so on) to be well defined and simple, namely, simply supported. The physical models developed regarded the bone as a SDOF system (one single bending mode) connected with mass-spring-dashpot elements representing interposing skin and vibrating muscle and soft tissue mass.

The problem of identifying (instead of assuming) the bone's own in vivo vibration modes, if even formulated,[7] remains unsolved. Furthermore, the question arises whether a simple beam model is adequate for a real bone. Finally, the proposed simply supported models cannot be used for the analysis of the dynamic response of bones to an arbitrary force input in different boundary conditions.

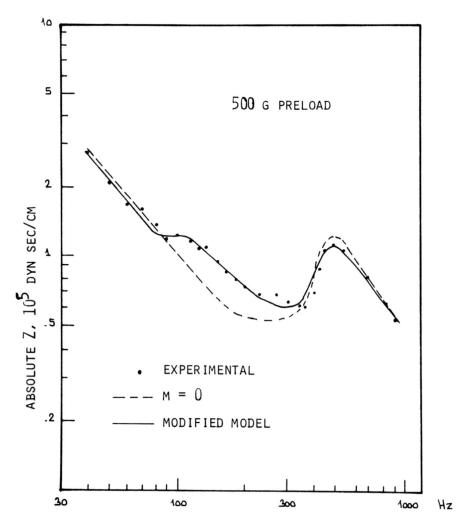

FIGURE 24. Measured and predicted ulnar impedance. M, muscle mass per unit length. M = 0 corresponds with the model of Figure 25; the modified model is the model of Figure 25, in which the beam representing the ulna is replaced by the ulna-musculature model of Figure 26. (From Orne, D. and Mandke, J., *J. Biomech.*, 8, 145, 1975. With permission.)

The development of finite element computing methods allowed to take into account the real geometry; mass, stiffness, and damping distribution of a bone; and to account for any boundary conditions. At the same time, the rapid progress in digital data processing techniques and the implementation of FFT routines, made it possible to determine frequency response functions using an impact or random noise signal as force input and thus to analyze a wide frequency range in one simple test. The development of "modal analysis" software allows the experimental determination of all modal parameters on excised as well as in vivo bones.

1. Finite Element Modeling

An important concept in dealing with bending stiffness of beams with nonsymmetrical sections (as bones) is the principal axis concept. Each cross section has two mutually perpendicular principal axes going through its centroid (Figure 27). The principal axes x and y are defined as axes which behave as the symmetry axes of a symmetrical beam cross

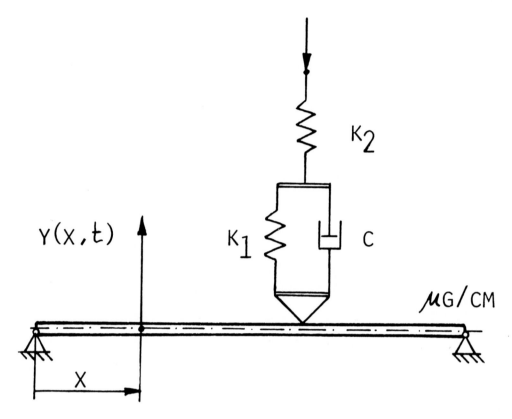

FIGURE 25.   Mechanical model for the ulna-interposing soft tissue system. The interposing soft tissue is represented by a tri-parameter system, characterized by the stiffness K_1, K_2, and the damping value C. μ, weight per unit length of ulna. (From Orne, D., *J. Biomech.*, 7, 252, 1974. With permission.)

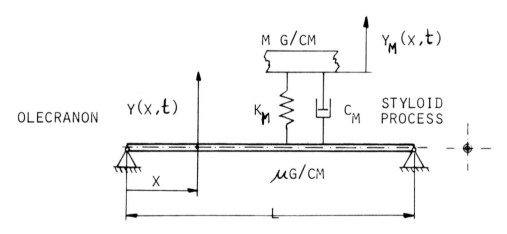

FIGURE 26.   Modified model for the ulna, including muscle mass. K_M, C_M, M, parameters of vibrating muscle mass per unit length. (From Orne, D. and Mandke, J., *J. Biomech.*, 8, 144, 1975. With permission.)

section. This means that (for a beam which is longitudinally uniform) if a bending moment is exerted along a principal axis, deflection is in the plane perpendicular to that axis (which is not the case for bending around an arbitrary axis). The static equilibrium relation between bending moment and radius of curvature in deflection is (for longitudinally uniform beams)

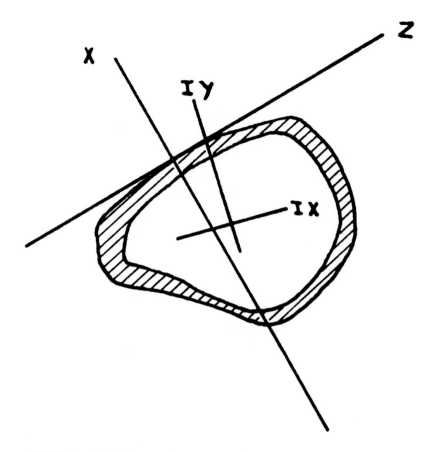

FIGURE 27.   Principal axes and measurement reference axes in a tibial cross section. IX, axis of minimum inertia (second area of cross section); IY, axis of maximum inertia (second area of cross section); x, normal to the medial face; z, tangential to the medial face.

$$1/R = \frac{M}{EI_{xx}} \qquad (98)$$

if M is a moment along the x axis and

$$1/R = \frac{M}{EI_{yy}} \qquad (99)$$

if M is a moment along the y axis, I is the second moment of area (inertial moment) about an axis through the centroid of cross section I varying with direction; I_{xx} and I_{yy} are the minimum and maximum values of I. Referring to Equation 53, the existence of two different principal directions and associated I values for a beam results in two different single bending modes and associated natural frequencies.

Extended structural analyses of human tibias, concerning among others, the variation in the principal axis directions (rotation or ''pretwist''), and associated I_{xx} and I_{yy} values along the shaft, were made by Minns et al.[18] and by Piziali et al.[19] The axis of minimum I is close to the sagittal direction; the axis of maximum I is close to the mediolateral direction.

Orne and Young[20] computed mode shapes and natural frequencies of a cantilevered canine radius by means of a linear elastic finite elements model. Variation of mass and geometry along the length of the bone as well as ''pretwist'' were taken into account, and the influence

Table 1
NATURAL FREQUENCIES FOR A FRESH EXCISED
HUMAN TIBIA CALCULATED USING A CONSISTENT
MASS FORMULATION

Model[a]	Single bending[b] I	Single bending[b] II	Double bending[b] I	Double bending[b] II	Axial displacement	Torsion
1. Free-free						
a	301.74	410.94	946.05	1225.09	2733.6	1364.4
b	301.94	409.99	948.88	1217.52	2733.6	1364.4
c	297.78	420.65	932.56	1216.85	2713.8	1352.1
d	297.89	419.89	937.90	1204.78	2710.6	1353.2
2. Hinged-hinged						
a	170.49	229.57	737.35	963.88	2009.0	1161.9
b	170.99	228.28	742.76	952.10	2009.0	1161.9
c	168.12	227.88	711.16	951.41	2032.0	1162.9
d	168.75	226.41	719.81	938.22	2014.4	1159.8
3. Fix-free						
a	91.78	132.19	467.20	619.28	2009.0	1161.9
b	92.06	131.33	468.96	615.60	2009.0	1161.9
c	91.41	131.26	462.94	613.63	2022.0	1155.4
d	91.69	130.45	464.67	609.89	2025.9	1153.9
4. Fix-fix						
a	555.52	730.38	1436.29	1840.69	3727.3	2836.5
b	573.90	708.32	1466.67	1805.19	3727.3	2836.5
c	579.65	727.39	1431.99	1825.40	3697.3	2837.7
d	601.18	701.85	1457.46	1795.76	3702.5	2833.1

[a]   a, straight without pretwist; b, straight with pretwist; c, curved without pretwist; d, curved with pretwist.
[b]   Bending I, about axis of EI_{min} bending II, about axis of EI_{max}.

(From Hight, T. K., Piziali, R. L., and Nagel, D. A., *J. Biomech.*, 13, 273, 1980. With permission.)

of shear deformation and rotatory inertia terms in the equation of motion was studied. Inclusion of fairly large pretwist angles (from -14 to $12°$), shear deformation, and rotatory inertia had only negligible influence on the first three frequencies of transverse vibration in either the lateral or cranial direction. The predominant influence was attributed to the variations in mass and flexural stiffness EI along the length of the bone. In order to get complete agreement between experiment and calculation, different values of E had to be prescribed for bending in the lateral direction and for bending in the cranial direction.

Hight et al.[21] computed the natural frequencies of a human tibia. First the hinged-hinged case was calculated using the consistent mass, lumped mass, and consistent mass + rotatory inertia + shear formulations. The results showed very little variation among the three matrix types for the same geometry. The further calculations were made using the consistent mass formulation and involved the comparison of standard-type boundary conditions and the variation of twist and curvature parameters. The results are shown in Table 1. Curvature and twist have only a slight influence except for particular modes in particular boundary conditions.

Finally, the authors quantified the changes in predicted natural frequencies which occur as the boundary conditions are varied through a spectrum of stiffness ranging from solidly fixed to free. As could be expected, a high sensitivity to boundary conditions was found. It is therefore essential to accurately estimate the stiffness of boundary "springs" in order to predict frequency response.

2. Modal Analysis

The modal analysis procedure as applied on human tibias by Van der Perre et al.[17] involves (see Figure 20) two functions.

Experimental determination of frequency response functions — Similar to the discretization procedure (see Section II.C.4), a number of points ($i = 1,n$) are selected on the tibia, such that the motion of the tibia is characterized adequately by the motion of these points. One excitation point (j) is selected, and now the acceleration frequency responses in all points i ($i = 1,n$) to a force input in point j are measured one after the other. This means n (acceleration measured in one direction), 2n (two-dimensional measurement), or 3n (three-dimensional) frequency response measurements. Each frequency response measurement involves:

1. Excitation of the tibia in point j by a force f(t) (Hammer impact was chosen for this application; f(t) is monitored by a load cell mounted in the hammer, see Figure 14)
2. Simultaneous measurement of acceleration a(t) in point i, using an accelerometer (see Figure 17)
3. "Low-pass filtering" (frequencies above the range of interest are filtered out) and analog to digital conversion (ADC) of both signals
4. FFT, resulting in $F(\omega)$ and $A(\omega)$, and computation of the acceleration frequency response function (transfer function TF) $A(\omega)/F(\omega)$

Determination of modal parameters by curve fitting — Fitting of Equation 72 to the n (2n or 3n) displacement frequency response functions $H_{ij}(\omega)$ obtained from the acceleration frequency response functions.

Force and acceleration signals are processed in a HP 5451C Fourier System provided with an analog to digital converter (ADC) and a 32K computer with disk memory and graphic terminal and implemented with a FFT routine with a keyboard program for the determination of frequency response functions (application of appropriate antileakage and antinoise windows, averaging) and with user programs for the curve fitting procedure. In all experiments, the response is analyzed up to a maximum frequency of 2500 Hz. The software used was developed by the Modal Analysis Groups of the Departments of Mechanical Engineering of the University of Leuven and the University of Cincinnati.[22]

The procedure results in the determination of the "modal parameters", i.e., for each mode k:

1. ω_{d_k}, damped natural angular frequency ($2\pi f_{d_k}$, f_{d_k}, cyclic frequency)
2. σ_k, damping value $\sigma_k = \omega_{d_k} \zeta_k$
3. The mode shape, indicated by the displacement vector (amplitude and phase) in each measurement point

The mode shapes are visualized in an animated display and can visually be identified, e.g., as rigid body, (single or multiple) bending, torsional modes, etc. In the cases where interaction between adjacent modes can be neglected, a mode shape k can be estimated graphically by taking the peak heights at ω_{d_k} in all measuring points and plotting them on the figure (Figure 28).

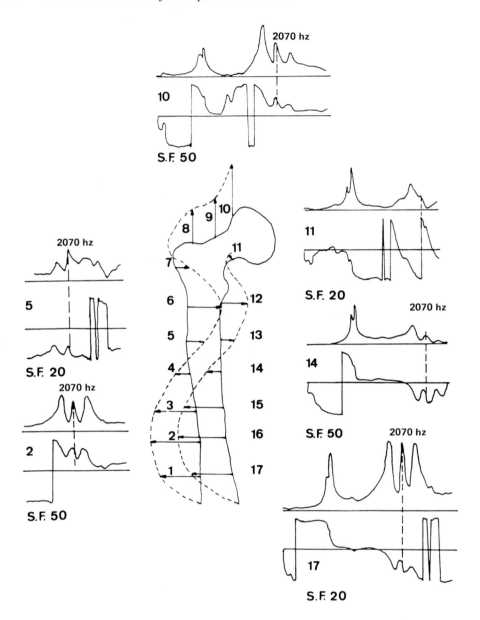

FIGURE 28.   Graphical reconstruction of mode shapes. The 2070-Hz mode shape of a dry human femur is obtained by plotting the 2070-Hz frequency response amplitude (taking the phase sign into account) in each point on the figure in the measurement direction.

a. Dynamic Characteristics of the Free Tibia

Eight dry tibias and two fresh tibias were analyzed in the free-free condition (Tables 2 and 3). A detailed three-dimensional modal analysis of one dry tibia (tibia 10) revealed two separate bending modes, corresponding to the two principal directions for bending (Figure 29), as was also predicted by the modal calculations of Hight et al.[21]

The direction of deflection associated with the first mode (minimum EI) is for most tibias nearby ($\pm 10°$ deviation) the direction normal to the medial face (x direction), while the deflection direction of the other mode is close to the direction tangent to the medial face (z direction). From the body reference axes, the first mode implies motion roughly in the mediolateral direction, whereas the second mode motion is roughly in the sagittal direction (see Figure 27).

Table 2
**NATURAL FREQUENCIES AND DAMPING RATIOS
FOR SINGLE BENDING MODES OF DRY EXCISED
HUMAN TIBIAS OBTAINED EXPERIMENTALLY BY
MODAL ANALYSIS**

	I, axis of minimum EI		II, axis of maximum EI	
Tibia nr.	Frequency[a] (Hz)	Damping ratio (%)	Frequency[a] (Hz)	Damping ratio (%)
2	424(X,Z)	0.8	659(Z)	1.0
3	594(X)	0.7	736(Z,X)	0.9
6	481(X)	0.9	680(Z)	0.9
8	522(X,Z)	0.8	778(Z,X)	1.4
9	690(X)	1.0	926(X,Z)	1.7
10	514(X,Z)	1.0	659(Z,X)	0.8
11	547(X)	0.4	881(Z)	0.9
12	540(X)	0.9	676(Z,X)	0.6

[a] X and Z, measuring directions in which each mode was observed.

(From Van der Perre, G., Van Audekercke, R., Vandecasteele, J., Martens, M., and Mulier, J. C., *Am. Soc. Mech. Eng. — Biomech. Symp.,* 1981, 173. With permission.)

Table 3
**NATURAL FREQUENCIES AND DAMPING RATIOS
FOR SINGLE BENDING MODES OF FRESH EXCISED
HUMAN TIBIAS OBTAINED EXPERIMENTALLY BY
MODAL ANALYSIS**

	I, axis of minimum EI		II, axis of maximum EI	
Tibia code	Frequency (Hz)	Damping ratio (%)	Frequency (Hz)	Damping ratio (%)
TIBHEMA	310(X,Z)	3.5	421(Z,X)	4.2
TIB6131	292(X)	4.2	412(Z)	5.4

[a] X and Z, measuring directions in which each mode was observed.

(From Van der Perre, G., Van Audekercke, R., Vandecasteele, J., Martens, M., and Mulier, J. C., *Am. Soc. Mech. Eng. — Biomech. Symp.,* 1981, 173. With permission.)

For the seven other tibias tested, single bending modes were determined by the following procedure: accelerations were measured in a row of seven to ten points along the medial face (see Figure 20) and excitation was by hammer impact in the middle of the tibia. For each tibia, two different tests were made (see Figure 27): (1) excitation and acceleration measurement perpendicular to the medial face (x direction) and (2) parallel to the medial face (z direction).

In Tables 2 (dry tibias) and 3 (fresh tibias), the obtained frequency and damping values are listed and the directions (x and z) in which each mode was observed are indicated. The natural frequency found for both fresh tibias were close to those calculated by Hight et al.[21] When compared with dry excised tibias, the fresh excised tibias show systematically lower values for the natural frequencies and higher values for the damping ratios.

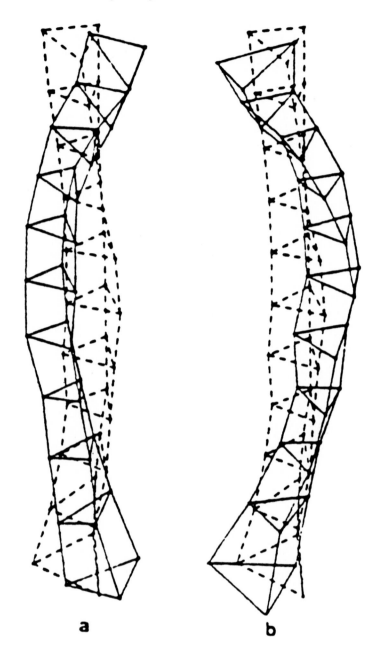

a **b**

FIGURE 29.   Single bending modes of a dry excised human tibia (free-free condition). (a) Bending mode associated with EI_{min} (roughly mediolateral motion); (b) bending mode associated with EI_{max} (roughly sagittal motion).

The presence of bone marrow in fresh tibias was simulated experimentally by injecting a viscous fluid into the medular canal of a dry tibia. From this experiment (Table 4) it appears that the bone marrow accounts for the major part of the frequency shift between dry and fresh tibias.

The application of modal analysis on free excised tibias is important from the following points of view: (1) it gives an understanding of the dynamic response characteristics of the bone as such, which is necessary for the understanding of its in vivo behavior, and (2) it

Table 4
EXPERIMENTAL SIMULATION OF BONE MARROW BY
INJECTING A VISCOUS FLUID INTO THE MEDULAR CANAL
OF A DRY TIBIA

Tibia	I, axis of minimum EI (Hz)	II, axis of maximum EI (Hz)
Tibia 5, dry	500	644
Tibia 5, simulated bone marrow	357	465

allows the evaluation of the various mathematical models which are proposed. More specifically, it makes it possible to specify the simplest possible mathematical model which can be used to accurately simulate the vibrational characteristics of long bones.

Using the above-mentioned modal analysis data, the applicability of a modified simple beam model to the free tibia was evaluated as follows.[23] For six tibias of Table 2, the E value was estimated by fitting a mathematical model to the modal analysis data. This mathematical model approaches the shaft by a consistent-mass finite element discretization procedure (representing it as a combination of eight short beams with constant E value) and represents the tibial ends as vibrating rigid bodies. The tibias were cut into blocks which were weighed and the I values of the end sections of each block were determined computergraphically. The advantage of determining E by fitting this model to the modal analysis data is that for each bone an individual, dynamic value of E is determined. After optimizing the E value, agreement between modal analysis and model calculated resonant frequencies and mode shapes for bending modes in both directions was very satisfying.

A modified simple beam model was then evaluated as follows. The "effective flexural rigidity" EI (E, the thus obtained value; I, an average of the measured I values over the shaft) was substituted in Equation 53 in order to obtain α as a "shape factor". The average value over the 6 tibias was $\alpha = 3.0155 \pm 0.233$ SD (whereas $\alpha = 3.56$ for the first free-free bending mode of a simple beam). In Table 5 the effective EI values obtained from the actual I distribution along the shaft are compared to the values obtained by using the above-determined mean value α in Equation 53.

$$\omega_k = 2\pi f_k = 2\pi\alpha \sqrt{\frac{EI}{\mu\ell^4}}$$

E, Young's modulus
I, second moment of area of the beam cross section $\Big\}$ EI, bending stiffness   (53)
μ, mass per unit length
ℓ, beam length

A similar comparison was made for the critical load (buckling load) P. The critical load is the value of the axial load at which a sideways deflected beam becomes unstable, i.e., is about to buckle. For a simply supported uniform beam (Figure 30), the critical load P is

$$P = \frac{\pi^2 EI}{\ell^2} \tag{100}$$

For the tibias of Table 5, critical loads calculated from the finite element model were compared with the critical loads obtained from Equation 100, in which EI values obtained from Equation 53 with $\alpha = 3.1055$ were substituted. Hence the latter critical loads were estimated from the natural frequencies by a simple beam approximation. Although the differences in "measured" and calculated EI values are within 10%, the uniform beam model should be improved by individualizing the shape factor α for each tibia (using X-ray photographs) or modeling the bone ends as additional vibrating masses.

Table 5
COMPARISON BETWEEN EI AND P VALUES CALCULATED
USING A SIMPLE-BEAM MODEL WITH α = 3.0155
("CALCULATED") AND EI AND P VALUES OBTAINED
FROM THE ACTUALLY MEASURED I DISTRIBUTION
ALONG THE SHAFT ("MEASURED")

Tibia[a]	ℓ (m)	m (kg)	f (Hz)	$(EI)_{calc}$ (Nm²)	$(EI)_{"meas"}$ (Nm²)	$(P)_{calc}$ (N)	$(P)_{"meas"}$ (N)
I	0.350	0.136	424	112.3	155.5	9023.6	8805.5
II			659	271.3	379.2	21858.2	20414.4
I	0.362	0.128	561	204.8	212.0	15424.6	15225.3
II			736	352.4	392.6	26541.0	23586.9
I	0.352	0.176	521	223.3	191.3	17787.0	10174.3
II			779	499.1	473.4	39755.9	32806.6
I	0.306	0.077	695	114.2	104.3	12037.1	10935.9
II			925	202.3	187.3	21323.2	19082.1
I	0.365	0.190	515	262.6	257.7	19454.0	16684.2
II			661	432.5	454.0	32040.6	24875.2
I	0.343	0.174	547	225.1	200.9	18883.7	17397.9
II			883	586.6	618.9	49210.0	51622.9

[a] I, first bending mode; II, second bending mode.

(From The University of Leuven and Hewlett-Packard, Proc. Int. Seminars of Leuven, 1978-1981.[22])

b. Modal Analysis of Human Tibias In Vivo

In a first study,[24] the modal parameters of the tibias of 20 men of age 22 ± 2 years were measured. For this purpose, frequency response measurements were made in eight points along the medial face, with the tibia hanging loosely with a knee angle of ± 90°. This study revealed a bending mode at 300 to 350 Hz (amplitude maximum in the acceleration spectrum*) which was interpreted as the first bending mode in the plane of minimum flexural rigidity.

In a more recent paper,[17] the same authors processed in vivo measured frequency response functions (which display high damping and strong coupling between adjacent modes) by an advanced curve fitting procedure, which resulted in the identification of two separate bending modes, corresponding with both principal directions (Table 6). The observed mode shapes indicate practically free-free boundary conditions, which is an important observation in connection with the influence of the joints in the test conditions used. The results as presented in Table 6 imply that the in vivo measured frequency responses are subject to a certain coupling (overlap) of adjacent modes (250 ↔ 400 Hz), in which the share of both modes is dependent upon the position of the measuring point and the direction of excitation and acceleration measurement. As the accelerometer is placed normally to the medial face, the first mode is always predominant, but the overlap with the second mode can affect the determination of the natural frequency of the first. Therefore, when designing a practical procedure for measuring tibial natural frequency measurements on patients, one should be aware of this effect. Such a procedure is now being developed by the authors. The damping ratio ζ, as determined from curve fitting, amounts for the first bending mode to 25 to 30%.

In order to verify these observations, and especially in order to evaluate the influence of

* Value to be multiplied by $(1 - \zeta^2)$ in order to obtain the damped natural frequency.

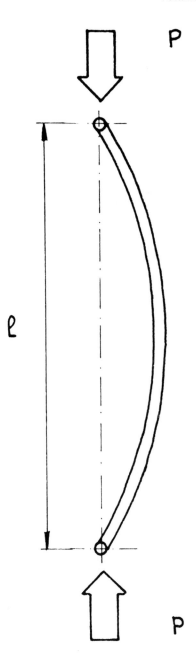

FIGURE 30.   Buckling of a beam under
axial loading. If P is larger than the critical
load value, the beam becomes unstable.

the in vivo medium upon the vibration behavior of the tibia, a systematical dissection was executed on an above-the-knee amputated underleg. The specimen was mounted with a knee angle of 90° and loosely hanging underleg, in order to simulate the in vivo measuring conditions. A complete modal analysis was made in six stages, as indicated in Table 7 (hammer excitation on the ankle, accelerometer in a row of eight points).

This led to the following results:

1.   The interpretation of in vivo excitated bending modes of the tibia (see Table 6) is

Table 6
**SINGLE BENDING MODES OF IN
VIVO HUMAN TIBIAS IDENTIFIED
EXPERIMENTALLY BY MODAL
ANALYSIS**

	Damped natural frequencies	
Subject	Single bending I[2] (Hz)	Single bending II[2] (Hz)
GVDP	280	350
JVDC	264	343

[a] I, bending about axis of minimum EI (clearly pronounced); II, bending about axis of maximum EI (weakly pronounced).

(From Van der Perre, G., Van Audekercke, R., Vandecasteele, J., Martens, M., and Mulier, J. C., *Am. Soc. Mech. Eng. Biomech. Symp.*, 1981, 173. With permission.)

Table 7
**SINGLE BENDING MODES OF AN
AMPUTATED UNDERLEG OBTAINED BY
EXPERIMENTAL MODAL ANALYSIS**

	Specimen state[b] (Hz)					
Bending mode[a]	1	2	3	4	5	6
Single bending I	264	267	314	—	320	318
Single bending II	336[c]	341[c]	—	420	—	435

[a] I, axis of minimum EI; II, axis of maximum EI.
[b] 1, intact specimen; 2, skin removed; 3, muscles removed; 4, ankle joint removed; 5, both joints removed (tibia-fibula with membrane); 6, free tibia.
[c] Component measured in x direction.

(From Van der Perre, G., Van Audekercke, R., Vandecasteele, J., Martens, M., and Mulier, J. C., *Am. Soc. Mech. Eng. Biomech. Symp.*, 1981, 173. With permission.)

confirmed. The amputation specimen displays the same modes, at 264 (clearly pronounced in the x direction) and 336 Hz (much less pronounced, while mainly z-directed), respectively. The evaluation of these modes was followed throughout the dissection experiment and led finally to the bending modes of the free tibia, at 318 and 435 Hz, respectively. The shape of the first bending mode (264 Hz) indicates free-free boundary conditions. The shape of the second mode (336 Hz) could not accurately be defined.

2. The skin, although affecting the measurements, is not consistently changing the natural frequencies, nor the mode shapes of the tibia.
3. As far as the first bending mode (x direction) is concerned, the frequency difference between the in vivo vibration and the free tibia is almost completely accounted for by the muscles. After removement of the muscles, the joints seem to play no considerable role. This is consistent with the observation that the in vivo vibration is free-free.

4.   Probably, the same holds for the second bending mode (z direction). Unfortunately, the frequency value immediately after removement of the muscles is missing, hence the possible influence of the ankle joint remains unknown.

C. Monitoring of Fracture Healing by Dynamic Response Analysis

If the degree of fracture healing could be assessed by a reliable, quantitative technique, this would most probably result in shorter immobilization periods. Furthermore, the same technique would be very useful in the early detection of nonunion as well as in the objective evaluation of techniques for artificial stimulation of healing (e.g., electrical stimulation). Measurements of ultrasound propagation speed[25] and ultrasound wave attenuation[26] through the fracture site have shown intrinsic and experimental limitations.

Bourgois and Burny[27] measured the rigidity of healing long bone fractures by a static mechanical test using strain gauges mounted on an external fixation rod. They are now extending their technique to internal fixation plates. This technique, although objective and reliable, is limited to fractures treated invasively.

Various teams have studied the application of dynamic response testing in fracture healing monitoring. A review of clinical measurements in connection with fracture healing is given in Table 8.

Markey and Jurist,[28] monitoring a 31-year-old woman with a comminuted right tibiofibular fracture, observed a "normal" acceleration frequency spectrum after 97 days, with a maximum at 191 Hz (instead of roughly 300 Hz[29].) Furthermore 24 subjects were measured on 70 occasions (mainly between 2 to 7 months). A significant correlation of $(f/f_o)^2$ with estimated strength (from X-ray evaluation) and time is propounded by the authors.

Sonstegard and Matthews[29] made time domain analyses of the acceleration signals in two points at both sides of the fracture and used as characteristic parameters (1, side of the excitation point; 2, opposite side): the amplitude ratio A_2/A_1, the slope ratio S_2/S_1, and the propagation velocity d/t (see Figure 31). The results obtained on seven dogs, the left radius of which was surgically fractured, are represented in Figure 32. In a study of 11 patients, 9 correct determinations of nonunion were made.

Steele and Gordon[30] developed a microprocessor controlled system ("SOBSA") measuring the transient response to a step in load, using a 40-Hz^2 square wave of excitation to a shaker. The displacement is obtained by two integrations of the acceleration signal, and $\alpha = P/W$ (P, buckling load; W, body weight) is calculated using a lumped mass model for the shaker, bone, skin, and muscle system. Results obtained on injured arms are listed in Table 9.

Doemland et al.,[31] in a paper on the development of a "microprocessor controlled system for the study of long bone mechanics", reported that data (frequency spectra of acceleration) have been collected on 38 subjects, without further details. The newly developed system is presently used for "longitudinal studies of fractured and healthy bones".

Cornelissen et al.[32] make use of a mathematical-physical model for a healing fractured bone, which is essentially a generalized version of the model proposed by Lewis.[33] The tibia is approximated by a simple beam. The fracture, whatever its real shape might be, is modeled by an "equivalent" straight zone of length D and (decreased) bending stiffness EI'. The position of the fracture is indicated by L_1/L_2 (Figure 33). Two characteristic parameters for the degree of fracture are defined.

1.   The "weakness index", $\delta = EI/EI'$, EI, effective bending stiffness of the intact, healthy bone; EI', bending stiffness of the fracture zone.
2. δ gives an indication of the healing degree, but is not an index of the overall strength of the fractured bone as a structural component of the fractured leg. As an index for the latter, the buckling strength P is used, i.e., the axial load at which a sideways deflected beam will collapse. In the case of a uniform Euler beam,

Table 8
REVIEW OF CLINICAL MEASUREMENTS IN CONNECTION WITH FRACTURE HEALING

Team, year	Excitation instrument + site + direction	Force input signal + measurement + processing	Vibration measurement + processing	Characteristic graph or quantities	Results in connection with fracture healing
1. Markey and Jurist, 1979	Modified loudspeaker driver (shaker) tibial tuberosity anterioposterior	Harmonic, variable frequency	Accelerometer strapped to medial malleolus	Acceleration vs. frequency, resonance peak	Correlation f^2/f_o^2 vs. X-ray estimated strength for 24 cases
2. Sonstegard and Matthews, 1976	Hammer with trigger supply on needle under $\pm45°$, 3—4 cm from prefracture needle	Impact	Accelerometers on 2 hypodermic needles at both fracture sides 6—15 cm apart	Acceleration vs. time on 2-channel storage oscilloscope Amplitude ratio shape ratio, propagation velocity (see Figure 31)	Surgical transverse fractures of the left radius of 7 dogs + 11 patients (see Figure 32)
3. Steele and Gordon, 1978	Impedance head, shaker	Static preload, 40-Hz wave monitored	Accelerometer, 2 hardware integrations → displacement	Transverse stiffness = displacement/force via model	10 arm injuries (see Table 9)
4. Doemland et al., 1980	Hammer	Impulse	Microphone via MPU → power spectrum	Resonant frequency	38 subjects
5. Cornelissen et al., 1982	Hammer impact on medial malleolus	Impact, monitored by load cell, Fourier transformed	Accelerometer in 5 points, Fourier transformed	Frequency response function	See Table 10

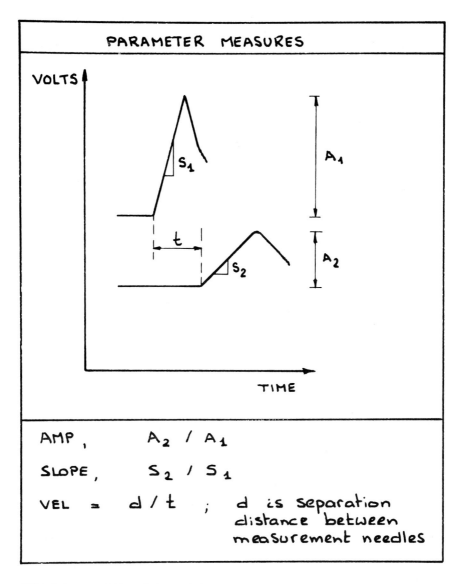

FIGURE 31. Definition of characteristic wave parameters. (1) Side of the excitation point; (2) opposite side. (From Sonstegard, D. A. and Matthews, L. S., *J. Biomech.*, 9, 691, 1976. With permission.)

$$P = \frac{\pi^2 EI}{\ell^2} \tag{100}$$

For any type of fracture (i.e., any value of L/D and L_1/L_2), graphs can be constructed relating the "dynamic ratio" (DR), f/f_o (f, cyclic natural frequency of fractured bone; f_o, cyclic natural frequency of the intact paired bone), to δ and to the relative buckling strength (RBS) P/P_o, the ratio of buckling strengths of fractured and intact bone. In Figure 34 RBS is plotted vs. DR for various L_1/L_2 values. It is important to note here that, while the relation δ-DR is very sensitive to the relative length of the fracture zone (L/D), this is not the case for the relation RBS-DR, which means that RBS and DR are influenced by the value of D in the same way. As precisely the determination of D is a rather difficult procedure which still has to be optimized using experimental models, so far only the estimation of the RBS P/P_o from the DR f/f_o is made.

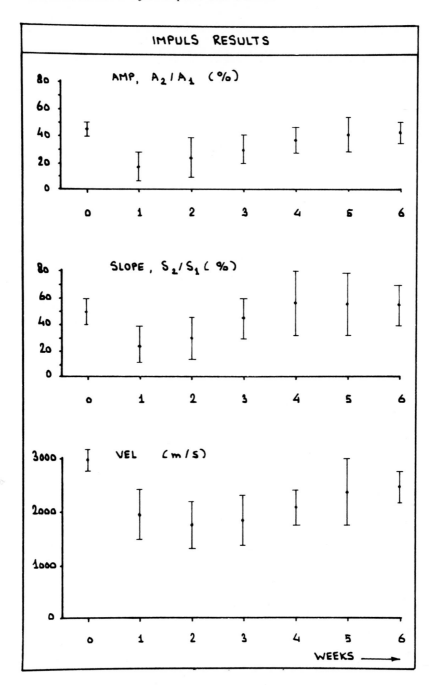

FIGURE 32.   Results (mean values and ranges) obtained on seven dogs the left radius of which was surgically fractured. (From Sonstegard, D. A. and Matthews, L. S., *J. Biomech.*, 9, 691, 1976. With permission.)

The model does not take into account influences of joints and soft tissues. However, (1) the in vivo vibration of the tibia is $\pm$ free-free and (2) the in vivo natural frequency f′ can be approximated by

Table 9
VALUES OF α = P/W (P, BUCKLING LOAD; W; BODY WEIGHT) OBTAINED ON INJURED ARMS BY THE "SOBSA" PROCEDURE

Subject	α, Uninjured	α, Injured	Comments
1	13.7	11.4	Wrist fracture, 1 year ago
2	15.7	5.1	Midshaft ulna fracture, 1 month ago
3	12.0	8.8	Hand fracture
4	11.0	12.0	Wrist fracture
5	12.3	8.9	Wrist fracture
6	15.8	14.3	Shoulder and arm fracture, 2×
7	13.5	14.8	Hand fracture, 3 months ago
8	14.5	6.1	Severe fracture of ulna, 2 years ago
9	14.7	5.7	Fused wrist, no grip, no use 4 years
10	12.8	8.2	Injured shoulder, 50% use for 3 years

(From Steele, C. R. and Gordon, A. F., in *Advances in Bioengineering,* ASME, 1978, 86. With permission.)

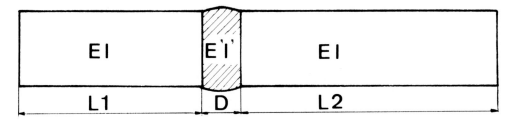

FIGURE 33.   Model for a healing fractured tibia EI, bending stiffness of intact bone; EI′, bending stiffness of fracture zone; D, equivalent length of fracture zone.

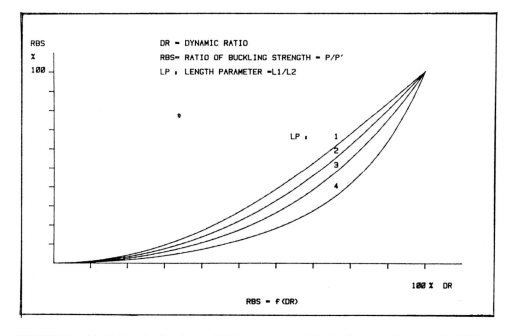

FIGURE 34.   Model for a healing fractured tibia: calculated graphs of relative buckling strength (RBS) vs. dynamic ratio f/f_o (DR); f, cyclic natural frequency of fractured bone; f_o, cyclic natural frequency of intact bone.

Table 10

VALUES OF DR (DYNAMIC RATIO) AND RBS (RELATIVE BUCKLING STRENGTH) OBTAINED FROM FREQUENCY RESPONSE MEASUREMENTS ON HEALING FRACTURED TIBIAS

Patient Days:	99	104	112	125[a]	132	160	182	196
DR =		73%	76%	80%	87%	86%		
RBS =		53%	58%	RFD 65%	76%	74%		
DR =	54%		68%					
RBS =	26%		43%					
DR =							75%	91%
RBS =							55%	RFD 82%

[a] RFD, removal of fixation day.

$$f' = kF\sqrt{1 - \zeta^2}$$

K, muscle mass factor (<1)

ζ, damping ratio

If K and ζ are assumed to be equal for the left and the right leg, then

$$\frac{f'/f'_o}{\text{(in vivo)}} = \frac{f/f_o}{\text{(free)}}$$

From frequency response measurements in five points, values of f' and f'_o (for the first bending mode) are obtained by modal analysis. The latter procedure is necessary to resolve the frequency response function into its various vibrational components. Fractures treated with conventional plaster casts as well as fractures treated with external fixators are monitored. Vibration analysis results from the latter are compared with the results of the Bourgois-Burny static test (Table 10).

D. Diagnosis of Osteoporosis

Osteoporosis is a loss of bone mass as a result of which the skeleton is eventually no longer capable to resist normal mechanical loading, and collapse of vertebrae or fracture of skeletal parts can occur. The development of a reliable noninvasive technique for the early detection of osteoporosis would mean an important step in health care.

As mentioned above, Jurist et al.[5,11] observed a significant difference in the value of FL (resonant frequency $\times$ length) measured on the ulna of a number of osteoporotic patients with respect to the value obtained on normal patients. However, as osteoporosis is essentially a decrease in bone mass, the simple-beam formula $f = \alpha\sqrt{EI/\mu\ell^4}$ indicates that, since both EI and μ decrease, the effect upon f is not easily predictable. Hence, μ and EI must be determined separately by fitting a mathematical model to a measured frequency response (e.g., mechanical impedance) graph.[14,15,20] Alternatively mass factors could be determined by other means, e.g., by bone mineral analysis (BMC).

Local disuse osteoporosis effects in arms were measured by Steele and Gordon[30] (see Table 9), who used a lumped mass model for the shaker-bone-skin-muscle system, and thus took mass and damping of bone and its surroundings into account. Local disuse osteoporosis is relatively easy to study since one can usually compare to the intact paired mate. The problem of general senile or postmenopausal osteoporosis detection, however, requires an absolute evaluation of bone properties. Hence, an improved operational model as discussed in Section III.B.2.a is required, which takes into account geometrical and mass characteristics as well as muscle and joint effects. Assuming that eventually EI (or E or P or $\alpha = P/W$) can be measured accurately, the detection of osteoporosis will consist in the observation of an "abnormal value" of EI (or E or P or $\alpha = P/W$). "Abnormal" then means (1) out of range of values which can be considered as normal for people of a distinct category (sex, age, weight, stature) or (2) critical from the viewpoint of functional loading criteria valid for the person in question. Criterion 1 implies the availability of data banks and statistical analysis; Criterion 2 implies a quantitative knowledge of loading conditions in various activities. As a conclusion we might say that the detection of osteoporosis by vibration analysis is a very challenging task.

IV. ANALYSIS OF THE DYNAMIC RESPONSE IN ACTUAL LOADING CONDITIONS

Two cases of dynamic loading are typical, namely, impact (car crash, sports accidents, etc.) and vibration (occupants response to vehicle vibration, hand-arm response to vibratory machinery). Many studies have been published concerning the response of the whole body and/or body parts in such situations. As far as the response of bones is concerned, the major part of published studies were on the skull and the (lumbar) spine.

However, dynamic loading is not restricted to impact or vibration situations. Even in normal walking, the lower limb is subjected to a high-frequency impulsive load at heel strike.[34,35] It would be interesting to study the relation between fatigue fracturing and dynamic response of leg bones.

Since dynamic response analysis starts with the analysis of the free vibration modal parameters, the finite elements and modal analysis studies mentioned under Section III.B (mainly on tibia) are useful for dynamic response purposes as well. Below we will briefly review dynamic analyses of femur, skull, and spine.

A. Dynamic Analysis of the Femur

Van Audekercke et al.[36] determined experimentally the modal parameters of dry and fresh excised human femurs in a study on the influence of an implanted total hip prosthesis upon the femur's dynamic behavior (see Section V). The results were in good agreement with those obtained by Khalil et al.[37] in a combined analytical and experimental study (Figure 35).

B. Dynamic Analysis of the Skull

Mathematical modeling of the physical processes leading to head injuries has received considerable alternation in recent years. For a brief account of injury mechanisms and mathematical modeling of head response due to impact loading, we refer to Khalil and Hubbard.[38] Khalil et al.[39] also made experimental modal analysis studies of the human skull.

Another application is the study of the dynamic characteristics of the skull in connection with the distortion of bone-conducted signals in audiometry.[40] Finally, analysis of frequency response of skull and temporal bone can be used as a diagnostic tool for fracture identification and localization.[41]

C. Dynamic Analysis of the Human Spine

Cramer et al.[42] presented a continuum model of the spine, subject to distributed inertial loading of the human torso. Although the inertial loading was regarded as time varying, the

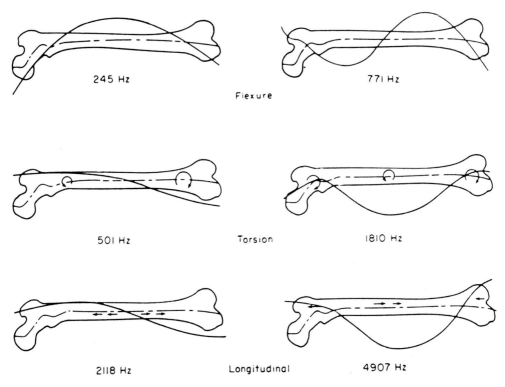

245 Hz

771 Hz

Flexure

501 Hz

Torsion

1810 Hz

2118 Hz

Longitudinal

4907 Hz

FIGURE 35.   Mode shapes of an embalmed human femur in the free-free condition. (From Khalil, T. B., Viano, D. C., and Taber, L. A., *J. Sound Vib.*, 75, 430, 1981. With permission.)

spine's response was treated by a quasistatic approach. Huijgens[43] calculated the bending vibration frequencies using simple beam theory. Pant and Tong[44] computed the normal modes of the vertebral column considering the 24 vertebras as a natural set of finite elements.

V. DYNAMIC ANALYSIS OF BONE IMPLANT ASSEMBLIES

A. Fracture Fixation Plates on Tibias

Van Audekercke et al.[36] studied the influence of metal fracture fixation plates on the dynamic behavior of intact human tibias. The idea was to identify differences with intact bones and in a further step to modify the plate design towards optimal "dynamic matching" between bone and implant.

Four dry tibias were analyzed by modal analysis before and after plate fixation. The results are listed in Table 11. The long plate fixation displays an interesting effect: the natural frequency of the single bending mode occurring in the direction normal to the plate increases, whereas the natural frequency corresponding to the other direction decreases. This phenomenon, if reproduced, could only be explained by the assumption that the screw joints allow small lateral displacements of the plate with respect to the bone. From this example, it appears that experimental dynamic analysis can be used as a technique for determining the properties of mechanical joints.

B. Total Hip Prostheses in Femurs

Modal analysis of intact femurs, femurs with cemented hip prosthesis, and femurs in which the latter was loose was made in view of (1) dynamic matching of the prosthesis to the femur and (2) development of a technique for the detection of loosening of hip prostheses.[36,45]

Table 11

**INFLUENCE OF A 10-HOLE FRACTURE FIXATION PLATE ON
NATURAL FREQUENCIES OF DRY EXCISED HUMAN TIBIAS**

	Tibia 3[a]	Tibia 10[a]	Tibia 12[a]	Tibia 6[a]
Single bending I				
Intact	561 (X)	515 (X,Z)	542 (X)	479 (X)
Plate midshaft	618 (X)	537 (Z)	611 (X,Z)	—
Plate distal	—	—	—	518 (X)
Single bending II				
Intact	736 (Z,X)	665 (X,Z)	676 (Z)	678 (Z)
Plate midshaft	686 (Z,X)	685 (X,Z)	656 (Z)	—
Plate distal	—	—	—	657 (Z,X)

[a]   X and Z, measuring directions in which each mode was observed.

Table 12

**MEASUREMENTS ON DRY HUMAN FEMURS WITH
AND WITHOUT IMPLANTED HIP PROSTHESIS**

a. Effect of Implanted Hip Prosthesis on Natural Frequencies

Femur No.	1[a]		2[a]		3[a]		4[a]	
Mode[b]	I	II	I	II	I	II	I	II
1	408	340	407	365	411	375	372	344
2	460	397	470	426	487	443	403	374
3	883	772	906	800	868	844	732	718
4	1153	1058	1207	1088	1136	1093	1036	966
5	1238	1140	1264	1164	1230	1166	1040	1033

b. Influence of Prosthesis Loosening on Natural Frequencies

Mode[b]	Fixed	Loose
1	375	300
2	410	380
3	775	640

[a]   I, intact femur; II, femur with implanted hip prosthesis.
[b]   1, first single bending mode; 2, second single bending mode; 3, torsional
mode; 4, first double bending mode; 5, second double bending mode.

An important problem in hip prosthesis implantation is the loosening of the bone cement fixation. In many cases of pain complaints, it is difficult to distinguish between two possible causes: low-grade infection or loosening of the prosthesis. In an early stage, this loosening cannot be detected by X-rays. The preliminary results of these measurements[45] can be summarized as follows (Table 12):

1.   The differences in natural frequencies among individual femurs are so large that mean values for intact bones cannot be used as a standard.
2.   Implanting a prosthesis causes frequency shifts. Especially the first torsional mode is sensitive ot this effect.

3.   When a prosthesis is cemented twice in the same bone, important frequency differences can occur. Hence, in order to test loosening by natural frequency testing, one will need reference values obtained by measuring immediately after implantation.
4.   In the case of a loose prosthesis, the following effects were observed:

● A slight decrease of natural frequencies for the single bending modes
● A neat decrease in natural frequency for the first torsional mode
● A rather "noisy" frequency response spectrum

The femur is difficult to be reached in vivo. Therefore, in clinical practice, measurements will need to be made using hypodermic needles brought into contact with the underlying bone and/or implant.

REFERENCES

1. **Clough, R. W. and Penzien, J.,** *Dynamics of Structures,* McGraw-Hill, New York, 1975.
2. **Broch, J. T.,** *Mechanical Vibration and Shock Measurements* (revised edition), Bruël and Kjaer, Eds., Naerum, Denmark, 1980.
3. **Burstein, A. H. and Frankel, V. H.,** A standard test for laboratory animal bone, *J. Biomech.,* 4, 155, 1971.
4. **Halvorsen, W. G. and Brown, D. L.,** Impulse technique for structural frequency response testing, *Sound Vib.,* 11, 8, 1977.
5. **Jurist, J. M.,** In vivo determination of the elastic response of bone. I. Method of ulnar resonant frequency determination, *Phys. Med. Biol.,* 15, 417, 1970.
6. **Jurist, J. M. and Dymond, A.,** Reproducibility of ulnar resonant frequency measurement, *Aerosp. Med.,* 41, 875, 1970.
7. **Doherty, W. P., Bovill, E. G., and Wilson, E. L.,** Evaluation of the use of resonant frequencies to characterize physical properties of human long bones, *J. Biomech.,* 7, 559, 1974.
8. **Jurist, J. M. and Kianian, K.,** Three models of the vibrating ulna, *J. Biomech.,* 6, 417, 1973.
9. **Spiegl, P. V. and Jurist, J. M.,** Prediction of ulnar resonant frequency, *J. Biomech.,* 8, 213, 1975.
10. **Cameron, J. R. and Sorenson, J.,** Measurement of bone mineral in vivo: an improved method, *Science,* 142, 5, 1963.
11. **Jurist, J. M., Hoeksema, H. D., Blacketter, D. A., Snider, R. K., and Garner, E. R.,** In vivo measurement of dynamic response of bone, in *Orthopaedic Mechanics, Procedures and Devices,* Vol. 3, Ghista, D. W. and Roaf, R., Eds., Academic Press, London, 1981, chap. 1.
12. **Campbell, J. and Jurist, J.,** Mechanical impedance of the femur: a preliminary report, *J. Biomech.,* 4, 319, 1971.
13. **Thompson, J. A.,** In vivo determination of bone properties from mechanical impedance measurements, in *Abstr. Aerosp. Med. Assoc. Annu. Sci. Meet.,* Las Vegas, 1973, 133.
14. **Orne, D.,** The in vivo driving point impedance of the human ulna — a viscoelastic beam model, *J. Biomech.,* 7, 249, 1974.
15. **Orne, D. and Mandke, J.,** The influence of musculature on the mechanical impedance of the human ulna — an in vivo simulated study, *J. Biomech.,* 8, 143, 1975.
16. **Saha, S. and Lakes, R. S.,** The effect of soft tissue on wave propagation and vibration tests for determining the in vivo properties of bone, *J. Biomech.,* 10, 393, 1977.
17. **Van der Perre, G., Van Audekercke, R., Vandecasteele, J., Martens, M., and Mulier, J. C.,** Identification of in vivo vibration modes of human tibiae by modal analysis, *Am. Soc. Mech. Eng.-Biomech. Symp.,* Van Buskirk, W. C. and Woo, S. L. Y., Eds., ASME, New York, 1981, 173.
18. **Minns, R. J., Bremble, G. R., and Campbell, J.,** The geometrical properties of the human tibia, *J. Biomech.,* 8, 253, 1975.
19. **Piziali, R. L., Hight, T. K., and Nagel, D. A.,** An extended structural analysis of long bones — application to the human tibia, *J. Biomech.,* 9, 695, 1976.
20. **Orne, D. and Young, D. R.,** The effects of variable mass and geometry, pretwist, shear deformation and rotary inertia on the resonant frequencies of intact long bones: a finite element model analysis, *J. Biomech.,* 9, 763, 1976.

21. **Hight, T. K., Piziali, R. L., and Nagel, D. A.,** Natural frequency analysis of a human tibia, *J. Biomech.,* 13, 139, 1980.
22. The University of Leuven and Hewlett-Packard, Modal analysis: theory, measurement techniques and industrial applications, in Proc. Int. Seminars, Leuven, Department of Mechanical Engineering, Van Brussel, H., Snoeys, R., and Peters, J., Eds., 1978-81.
23. **Cornelissen, Ph., Van der Perre, G., and Van Audekercke, R.,** Mathematical models for the vibrating tibia, paper presented at the 3rd Meet. Eur. Soc. Biomech., Nijmegen, January 1982.
24. **Vandecasteele, J., Van der Perre, G., Van Audekercke, R., and Martens, M.,** Evaluation of bone strength and integrity by vibration methods, in *Mechanical Factors and the Skeleton,* Stokes, I., Ed., John Libbey, London, 1980, 98.
25. **Abendschein, W. and Hyatt, G.,** Ultrasonic and physical properties of healing bone, *J. Trauma,* 12, 297, 1972.
26. **Brown, S. and Mayor, M. B.,** Ultrasonic assessment of early fracture callus formation, *Biomed. Eng.,* 11, 124, 1976.
27. **Bourgois, R. and Burny, F.,** Measurement of the stiffness of the fracture callus in vivo: a theoretical study, *J. Biomech.,* 5, 85, 1972.
28. **Markey, E. and Jurist, J. M.,** Tibial resonant frequency measurements as an index of the strength of fracture union, *Wis. Med. J.,* 73, 62, 1974.
29. **Sonstegard, D. A. and Matthews, L. S.,** Sonic diagnosis of fracture healing — a preliminary study, *J. Biomech.,* 9, 689, 1976.
30. **Steele, C. R. and Gordon, A. F.,** Preliminary clinical results using "SOBSA" for noninvasive determination of ulna bending stiffness, in *Advances in Bioengineering,* Am. Soc. Mech. Eng. Symp., ASME, New York, 1978, 85.
31. **Surber, J. L., Kitchen, K. G., Doemland, H. H., Ketzner, J. J., and Stanley, D. H.,** An application of microprocessor control to the study of long bone mechanics, *Biomed. Sci. Instrum.,* 15, 31, 1979.
32. **Cornelissen, M., Cornelissen, Ph., and Van der Perre, G.,** A dynamic model for a healing fractured tibia, in *Biomechanics: Principles and Applications (Developments in Biomechanics),* Vol. 1, Huiskes, R., Van Campen, D. H., and de Wijn, J. R., Eds., Martinus Nijhoff, The Hague, Netherlands, 1982, 213.
33. **Lewis, J.,** A dynamic model of a healing fractured long bone, *J. Biomech.,* 8, 17, 1975.
34. **Light, L. H., McLellan, G. E., and Klenerman, L.,** Skeletal transients on heel strike in normal walking with different footwear, *J. Biomech.,* 13, 477, 1980.
35. **Simon, S. R., Paul, I. L., Mansour, J., Muuro, M., Abernethy, P. J., and Radin, E. L.,** Peak dynamic force in human gait, *J. Biomech.,* 14, 817, 1981.
36. **Van Audekercke, R., Van der Perre, G., and Martens, M.,** Analysis of the dynamic behaviour of long bones and orthopaedic implants, paper presented at the 1st World Biomaterials Congr., Baden near Vienna, April 1980.
37. **Khalil, T. B., Viano, D. C., and Taber, L. A.,** Vibrational characteristics of the embalmed human femur, *J. Sound Vib.,* 75, 417, 1981.
38. **Khalil, T. B. and Hubbard, R. P.,** Parametric study of head response by finite element modeling, *J. Biomech.,* 10, 119, 1977.
39. **Khalil, T. B., Viano, D. C., and Smith, D. L.,** Experimental analysis of the vibrational characteristics of the human skull, *J. Sound Vib.,* 63, 351, 1979.
40. **Arlinger, S. D., Kyl'en, P., and Hellqvist, H.,** Skull distortion of bone conducted signals, *Acta Oto Laryngol.,* 85, 318, 1978.
41. **Christmann, C. and Holzweissig, F.,** On the estimation of the vibration of human skulls (German), *Anat. Anz.,* 142, 229, 1978.
42. **Cramer, H. J., King Lin, Y., and von Rosenberg, D. U.,** A distributed parameter model of the inertially loaded human spine, *J. Biomech.,* 9, 115, 1976.
43. **Huijgens, J. M. M.,** Bending vibrations in the human vertebral column, *J. Biomech.,* 10, 443, 1977.
44. **Pant, M. M. and Tong, B. Y.,** Normal modes of vibration of the vertebral column, paper presented at the New England Bioengineering Conf., 1977.
45. **Belien, H. and Goossens, L.,** Study of the Dynamic Behaviour of Bone-Implant Structures, Master of Engineering thesis, Katholieke Universiteit, Leuven, 1981 (in Dutch).

INDEX

A

Absorptiometry, 134
Acceleration frequency response function, 113
Accelerometers, 129, 130
Accuracy of models, 94
Acoustic emission, 20—21
Age-dependent changes in bone collagen, 15
Aging, 12—15, 38
β-Aminoproprionitrile (BAPN), 13
Anelastic deformation of bone, 16
Angular deflection, 80
Angular frequency, 141
Angular twist, 82, 85
Animal experimentation, 4, 6
Animal models, 6
Anisotropy, 94
  cortical bone, 10, 43
Assumptions of models, 94
Attenuation of ultrasound waves, 149
Axial direct stresses, 86, 87
Axial forces, 85
Axial loading, 11
Axial rigidity, 75
Axis
  of inertia, 81
  orientation of, 39
  osteon, 25
  principal, 81

B

BAPN, see β-Aminoproprionitrile
Bar model, 74—76, 85
Beam model, 13, 74, 78—82, 85, 115, 118, 145
  bending in, 80
  calculations in, 134
  cantilever in, 58, 84
  vibration in, 116
Beam-on-elastic-foundation theory, 74
Behavior, see also specific behaviors
  bending, see Bending
  compressive, see Compression
  deformational, see Deformation
  elastic, see Elasticity
  functional, 2
  linear elastic, 93
  plastic, 10, 66
Bending, 78
  of beams, 80
  mathematical model for, 11
  moments of, 85
  stiffness of, 118
  of tibia, 147
Bending strength, see also Flexural rigidity, 45
  of bone tissue, 11
Biaxial strength, 46

Biocompatibility, defined, 3
Bioglasses, 5
BMC, see Bone mineral analysis
Bone collagen age-dependent changes, 15
Bone mineral analysis (BMC), 154
Bone remodeling, 14
Bone tissue bending strength, 11
Boundary conditions, 67, 88, 141
  kinematic, 89
Boundary models, 134
Breakage of fiber, 19
Buckling (critical) load, 145
Buckling strength, 45, 149, 151

C

Canine radius, 139
Cantilever beam, 58, 84
Cement lines, 38
Ceramics, 4
Characteristic equation, 118
Characteristic wave parameters, 151
Circular shaft torsion, 82—85
Clinical evaluation, 6
Closed-form model, 68, 72—88, 91, 93
Closed loop research outline, 4
Coatings, 5
Collagen
  age-dependet changes in, 15
  cross-linking of, 14
  structures of, 66
Combined loading of slender bodies, 85—95
Compact bone elastic and ultimate properties, 45
Complex frequency response function, 108
Compliance, 113
Composite materials, 4, 18, 19, 21
Composition of bone, 12—15
Compound-bar theory, 76—78
Compound-beam theory, 74
Compressibility constant, 66
Compression
  fatigue damage due to, 26
  strength under, 45
  tension and, 10
Computers, 88, 132, 149
Concentrations of stress, 72, 75, 91, 92
Consistent mass, 140
Constitutive equations, 65, 66, 68, 70
  elastic properties of, 63
Continuum theory, 60, 66, 67
Convergence test, 91
Convolution theorem, 108
Coordinates, 124—126
Cortical bone
  anisotropy of, 10, 43
  compressive behavior of, 10
  failure mechanisms in, 21

maturation of, 39
   stress-strain characteristics of, 15
   trabecular bone vs., 66
Cracks, see also Microcracking
   fatigue, 17
   formation of, 21
   oblique, 24, 33
Creep, 25, 32
Critical (buckling) load, 145
Critical damping value, 112
Cross-linking of collagen, 14
Cross-sectional dimensions, 136
Curve fitting, 141
Cutting principle, 57

D

Damped impulse response, 115
Damped natural frequency, 112, 121
Damping, 100, 102, 115, 141
   critical, 112
   of free vibration, 113
   influence coefficients of, 121
   nonproportional, 122—123
   proportional, 122—123
   ratios of, 143, 146
   tibia, 143
Debonding in composite materials, 19
Decalcification
   osteon, 32
   progressive surface, 12
Decay, 112
Decomposition of forces, 54, 58
Deflection, 101
   angular, 80
   shape of, 116
Deformation, 59, 60, 68
   anelastic, 16
   local state of, 63
   plastic, 10
   strains and, 96
Degree of mineralization, 12
Degrees of freedom, 115—116
   multiple, see Multiple degrees of freedom
   single, see Single degrees of freedom
Density, 136
   mesh, 91
Design evaluation, 67
Design optimization, 67
Development of materials, 4—5
Diagnosis of osteoporosis, 154—155
Diet, 13
Dimensions, 3
   cross-sectional, 136
Direct frequency response superposition concept,
   126
Direct strains, 62
Direct stress, 60
   axial, 86, 87
Discretization, 116, 126, 145

Displacement frequency response function, 113
Displacements
   nodal point, 92
   response to, 120
   static equilibrium, 100
   vector of, 126
Distributed parameter systems, 115
Distribution of internal load, 59
Driving point impedance test, 134—136
Dynamic analysis
   of femur, 155
   of long bones, 136—149
   of skull, 155
   of spine, 155—156
   static analysis vs., 100—102
Dynamic compliance, 113
Dynamic equilibrium, 117
Dynamic loading, 155
Dynamic magnification factor, 112
Dynamic ratio, 151
Dynamic response, 107—110
   analysis of, 132
   free vibration motion and, 102—115
   MDOF sytems, 118—126
   system, 119
Dynamics, 52
Dynamic stiffness matrix, 121

E

Effective flexural rigidity, 145
Elasticity, 66, 102, 117
   compact bone, 45
   constitutive equations, 63
   linear, 85, 93
   materials, 65, 67
   modulus of, see Modulus of elasticity
Elastic-perfectly plastic model, 11
Electrical strain gauges, 69
Electromagnetic shaker, 126, 127
Element
   finite, 88—95, 126, 127—141
   stiffness mtrix of, 90
Element mesh, 94
   of models, 89
Elongation, 63
Emission, 20—21
End effects, 75—77
Energy, 90
Equation of motion, 26
Equilibrium, 102
   conditions of, 54, 55, 58
   displacement of, 100
   dynamic, 117
   static, 100
Equivalent stress, 67
Evaluation
   clinical, 6
   design, 67
Experimentation

animal, 4, 6
  stress analysis, 72
Exponentially decaying window, 112
External forces, 102
  on body, 53
  vector of, 124

F

Failure
  cortical bone and, 21
  criterion for, 44
  multidirectional, 44—46
Fatigue
  bone and, 17—20
  compression, 26
  cracks from, 17
  flexural, 29
  limit on, 4
  strength of, 67
  tests of, 19
  uniaxial, 19
Feedback from senses, 100
FEM, see Finite element methods
Femoral neck fractures, 134
Femur
  dynamic analysis of, 155
  total hip prostheses in, 156—158
Fiber breakage in composite materials, 19
Filtering, 141
Finite element model, 88—95, 126, 127—141
Fissures, see Cracks
Fixation materials, 5
Fixation plates, 156
Flexural fatigue, 29
Flexural rigidity, see also Bending strength, 80
  effective, 145
  minimum, 146
Force analysis, 55, 57
  biomechanics, 55
Force function, 108
Forces, see also specific forces, 95
  axial, 85
  decomposition of, 54, 58
  elastic, see Elasticity
  external, see External forces
  gait, 53
  inertial, see Inertia
  internal, 76
  model of, 55
  moment of, 54
  transducers of, 130, 131
  transverse, 78, 81, 85
Fourier convolution theorem, 108
Fracture fixation plates on tibias, 156
Fracture healing, 100, 132, 149—154
  femoral neck, 134
Free body diagrams 54, 55, 57—59, 67
Freedom degrees, see Degrees of freedom
Free tibia, 142—145

Free vibration, 100, 104
  dynamic response and, 102—115
  MDOF system, 118
  parameters of, 110—112, 118—126
  subcritically damped, 113
  undamped, 102—103, 116—118
Frequency
  angular, 141
  damped, 112, 121
  MDOF systems, 120—122
  natural, see Natural frequency
  resonant, 133—134
Frequency domain, 109, 126
Frequency response, 104—107, 110—112
  acceleration, 113
  complex, 108
  displacement, 113
  matrix of, 120
  SDOF, 112—114
  superposition of, 126
  testing of, 126—132
Friction, 3
  stress of, 16
Fringe
  Moire, 68
  photoelastic, 73
Functional behavior, defined, 2
Functionality, 3

G

Gait
  cycle of, 42
  forces of, 53
Gauges, 41, 70, 149
  strain, 69
Generalized coordinate model, 126
Generalized mass, 103, 118
Generalized stiffness, 103, 118
Geometry, 88, 95—96
Gradients of stress, 91
Growth, 19

H

Hammer, 128
Harmonic motion, 102, 110
Haversian microstructure, 10
Head, 130, 131
Healing of fractures, see Fracture healing
Heel strike, 42, 155
Hip prostheses, 156—158
Hip replacement, 2, 156—158
Hooke's law, 63
Hydroxyapatite, 5

I

Impact, 155

Impedance
 driving point, 134—136
 mechanical, 114
 tests of, 134—136
 ulnar, 136
Impedance head, 130, 131
Implant testing, 3—4
Impulse response, 100, 104, 107—110
 damped, 115
 point, 120—122
Impulsive load, 155
Inertance, 113
Inertia, 102, 117
 polar moment of, 83
 principal axes of, 81
 rotatory, 140
 static (second) moment of, 79, 81
Infection, see also specific types
 low-grade, 157
Influence
 damping, 121
 mass, 117
 stiffness, 117
Instrumented hammer, 128
Interface
 conditions of, 67
 shear stess of, 77
 stresses at, 77, 92
Interface conditions, 88
Internal force, 76
Internal load, 57
 distribution of, 59
In vitro testing, 5
In vivo testing, 100
Isotropy, 40
 transverse, 44
Iteration procedures, 93

K

Kinematics, 52, 53
 boundary conditions of, 89
 of restraints 57

L

Laplace transformation, 114—115
Linear elasticity, 85, 93
Load, 96
 axial, 11
 conitions of, 88
 critical (buckling), 145
 distribution of, 59
 dynamic, 155
 impulsive, 155
 internal, see Internal load
 slender bodies, 85—95
 tensile, 34
 transmission of, 77

Local deformation, 63
Long bone MDOF dynamic analysis, 136—149
Longitudinal modulus, 10
Longitudinal splitting, 24
Long-term functionality, 3
Loosening of prosthesis, 157
Low-grade infection, 157
Low-pass filtering, 141
Lumped mass, 140, 149, 154
Lumped parameter model, 126

M

Magnification factor, 112
Mass, 100, 103, 115
 consistent, 140
 generalized, 103, 11
 lumped, 140, 149, 154
 matrix of, 117, 120
Mass influence coefficients, 117
Materials
 composite, 4, 18, 19, 21
 development of, 4—5
 elastic properties of, 65, 67
 fixation, 5
 mechanical properties of, 63—67
 properties of, 67, 88, 96
 surface-coated, 5
Mathematical models, 71
 for bending behavior, 11
Mathematical optimization, 55
Matrix
 frequency reponse, 120
 mass, 117, 120
 stiffness, see Stiffness matrix
 transfer, 120
Maturation of cortical bone, 39
MDOF, see Multiple degrees of freedom
Measurements, see also specific types
 steady-state response, 126
 ultrasonic, 38
Mechanical impedance, 114
Mechanical properties, 3
 of materials, 63—67
Mechanics
 rigid body, 52—59
 solid, 52, 59
Mesh
 density of, 91
 element, 89, 94
Metals, 4
Metric system, 95
Microcracking, 24
 in composite materials, 19
Microdamage, 10
Micromechanical events at yield, 18
Micromechanical studies, 16
Microprocessor controlled systems, 149
Microstructure, 10
Microyield phenomena, 15—17

Mineral analysis, 154
Mineralization, 14
  degree of, 12
Minimal potential energy, 90
Minimum fleural rigidity, 146
Modal analysis, 122, 137, 141—149
  of human tibias, 146—149
Modal coordinates, 124—126
Modal parameters, 118
  free vibration, 118—126
Modal superposition method, 126
Models, see also specific models, 67, 68, 126
  accuracy of, 94
  animal, 6
  assumptions in, 94
  beam, see Beam model
  boundary, 134
  elastic-perfectly plastic, 11
  element mesh of, 89
  finite element, 88—95, 137—141
  of forces, 55
  generalized coordinate, 126
  isotropic, 40
  lumped mass, 149, 154
  lumped parameter, 126
  mathematical, 11, 71
  three-D, 91
  validity of, 71, 85, 88, 94
Modes, see also specific types
  bending, 147
  real, 123
Mode shapes, 118, 139, 141
  normalized, 125
  vector of, 117
Modulus
  elastic, see Modulus of elasticity
  longitudinal, 10
  of rupture, 11
  shear, 63, 85
  Young's, see Young's modulus
Modulus of elasticity, see also Young's modulus, 4,
    14, 38, 63, 136
Moire fringe techniques, 68
Moment of inertia
  polar, 83
  static (second), 79, 81
Moments
  bending, 85
  of force, 54
Motions
  equation of, 126
  free vibration, see Free vibration
  harmonic, 102, 110
Multidirectional failure, 44—46
Multidisciplinary teams, 7
Multiple degrees of freedom (MDOF) systems,
    115—126
  dynamic response of,118—126
  free vibration of, 118
  frequency characteristics of, 120—122
  long bones, 136—140

Muscles, 148

N

Natural frequency, 103, 118, 139
  angular, 141
  damped, 112, 121
  of tibia, 140, 143
Necrosis and resorption, 67
Neutral line, 79
Neutral plane, 81
Nodal point displacement, 92
Nonhomogeneity, 94
Nonproportional damping, 122—123
Normal coordinates, 125
Normalized mode shapes, 125

O

Oblique cracking, 24, 33
Off-axis strength, 47, 48
Optimization
  design, 67
  mathematical, 55
Orientation, 10
  of axes, 39
Orthogonality properties, 125
Osteoblastic activity, 38
Osteons, 14, 16, 24
  axis of, 25
  decalcified, 32
  primary, 38
  secondary, 38
  single, 25
  strain of, 16
Osteoporosis, 100, 132, 134
  diagnosis of, 154—155

P

Pain, 157
  threshold of, 135
Parametric analysis, 67, 91, 93, 95
Photoelasticity
  analysis of, 72
  fringes of, 73
  three-D, 68, 69
Photon absorptiometry, 134
Plane
  neutral, 81
  stress on, 63, 69
Plastic behavior, 10, 66
Plate and shell theories, 74
Plates of fracture fixation, 156
Point implse response, 120—122
Poisson's ratio, 40, 63, 66, 70, 93
Polar moment of inertia, 83
Polyethylene, 2

Porous coatings, 5
Potential energy, 90
Pretwist, 139
Primary osteons, 38
Principal axes of inertia, 81
Principal strains, 41
Principal stresses, 41, 43
Principle of minimal potential energy, 90
Principle of Saint-Venant, 75
Progressive surface decalcification, 12
Propagation of ultrasound, 132
    speed of, 149
Proportional damping, 122—123
Proportionality constant, 118
Prostheses, see also specfic types
    loosening of, 157
    total hip, 156—158
Push-off, 42

Q

Quasiinertial response, 112
Quasistatic response, 112
Quasistatic tensile testing, 40

R

Rachitogenic diet, 13
Radius, 139
RBS, see Relative buckling strength
Real displacement vector, 126
Real modes, 123
Relative buckling strength (RBS), 151
Remodeling of bone, 14
Research outline, 4
Resonant frequency
    determinations of, 133—134
    of ulna, 133
Resorption, 38
    stress-related necrosis and, 67
Response
    displacement, 120
    dynamic, see Dynamic response
    frequency, see Frequency response
    impulse, see Impulse response
    quasiinertial, 112
    quasistatic, 112
    steady-state, 126
Restraints, 57
Rigid body, mechanics of, 52, 53—59
Rigidity
    axial, 75
    effective, 145
    flexural, see Flexural rigidity
    minimum, 146
    torsional, 85
Rosette, 69
Rotatory inertia, 140
Rupture modulus, 11

S

Saint-Venant principle, 75
Saint-Venant's warping theory, 85
SDOF, see Single degree of freedom
Secondary osteons, 38
Second (static) moment of inertia, 79, 81
Sensory feedback, 100
Shafts
    circular, 82—85
    torsion, 74
Shakers, 126, 127
Shape, 145
    deflection, 116
    mode, see Mode shape
    normalized, 125
Shear formulations, 140
Shear modulus, 63, 85
Shear slippage, 25
Shear strain, 62
Shear stress, 60, 77, 85
    components of, 60
    interface, 77
SI metric system, 95
Single degree of freedom (SDOF), 103—104
    frequency response function of, 112—114
Single osteons, 25
Skin, 148
Skull, 155
Slender body combined loading, 85—95
Slippage, 25
SMI, see Static moment of inertia
Softening of strain, 15
Soft tissue related effects, 135
Solid mechanics, 52, 59
Specimen preparation, 11
Speed of ultrasound propagation, 149
Spine, 155—156
Splitting, 24
Spring constant, 101
Standing waves, 123
Static vs. dynamic analysis, 100—102
Static equilibrium displacement, 100
Static (second) moment of inertia (SMI), 79, 81
Statics, 52
Steady-state response measurements, 126
Steady-state vibration, 106, 133—136
Sterilization,
Stiffness, 100, 103, 115
    bending, 118
    generalized, 103, 118
    influence coefficients of, 117
Stiffness matrix, 42, 117, 120
    dynamic, 121
    element, 90
Strain, 59, 60, 63
    analysis of, 70
    deformations and, 96
    direct, 62
    osteon, 16
    principal, 41

shear, 62
softening of, 15
stress and, see Stress-strain relationships
tissue, 13
ultimate, 14
uniaxial, 69
Strain gauges, 41, 70, 149
electrical, 69
Strain rosette, 69
Strength
bending, see Bending strength
biaxial, 46
buckling, 149
compressive, 45
fatigue, 67
off-axis, 47, 48
relative buckling, 151
tensile, see Tensile strength
ultimate, 65
yield, 14
Stress, 59, 60, 95, 96
at interface, 92
axial direct, 86, 87
components of, 61
concentrations of, 72, 75, 91, 92
direct, see Direct stress
equivalent, 67
friction, 16
gradients of, 91
interface shear, 77
necrosis and resorption related to, 67
plane, 63, 69
principal, 41, 43
shear, see Shear stress
three-D, 61, 62
two-D, 61
uniaxial, 63, 67, 69
Stress analysis, 41, 66, 68, 70
closed-form, see Closed-form model
experimental, 72
Stress-strain relationships, 10—12
in cortical bone, 15
curve of, 65
Strike of heel, 42, 155
Structural analysis, 95
of human tibia, 139
Subcritically damped free vibration, 113
Superposition concept, 126
Surface-coated materials, 5
Surface decalcification, 12

T

Teams, 7
Tensile loading, 34
Tensile strength, 14
off-axis, 48
Tensile test, 63
quasistatic, 40
Tension

copression and, 10
Testing
biomechanical, 5—6
convergence, 91
driving point impedance, 134—136
fatigue, 19
frequency response, 126—132
impedance, 134—136
implants, 3—4
in vitro, 5
in vivo, 100
quasistatic, 40
tensile, see Tensile test
uniaxial fatigue, 19
THR, see Total hip replacement
Three-D model, 91
Three-D photoelasticity, 68, 69
Three-D stress state, 61, 62
Tibia, 141, 153
bending modes of, 147
damping ratios of, 143
fracture fixation plates on, 156
free, 142—145
modal analysis of, 146—149
natural frequency of, 140, 143
structural analysis of, 139
Time domain-frequency domain relation, 109
Tissue, see also specific types
bone, 11
soft, 135
strain of, 13
Torsion
of circular shafts, 82—85
theory of, 74, 82
Torsional creep, 25, 32
Torsional rigidity, 85
Total hip replacement (THR), 2, 156—158
Toughness, 65
Trabecular bone vs. cortical bone, 66
Transducers, 130, 131
Transfer
function of, 114—115
matrix of, 120
Transverse forces, 78, 81, 85
Transverse isotropy, 44
Tricalciumphosphates, 5
Twisting, 82, 85
Two-D stress state, 61

U

UHMWPE, see Ultrahigh molecular weight
polyethylene
Ulna, 134
impedance of, 136
resonant frequency of, 133
Ultimate strain, 14
Ultimate strength, 45, 65, 149, 151
Ultrahigh molecular weight polyethylene
(UHMWPE), 2

Ultrasonic measurements, 38
Ultrasound propagation, 132
  speed of, 149
Ultrasound wave attenuation, 149
Undamped free vibration, 102—103, 116—118
Uniaxial fatigue tests, 19
Uniaxial strain, 69
Uniaxil stress, 63, 67, 69

V

Validity of models, 71, 85, 88, 94
Vibration, 155
  beam, 116
  free, see Free vibration
  parameters of, 110—112, 118—126
  steady-state, 106, 133—136
  subcritically damped, 113
  undamped, 102—103, 116—118

Viscoelasticity of bone, 10
Void growth in composite materials, 19

W

Warping theory, 85
Wave attenuation, 149
Wave parameters, 151
Weakness index, 149
Wear, 3
Window, 112

Y

Yield
  criterion for, 13
  micromechanical events at, 18
  strength of, 14
Young's modulus, see also Modulus of elasticity,
    63, 65, 66, 70